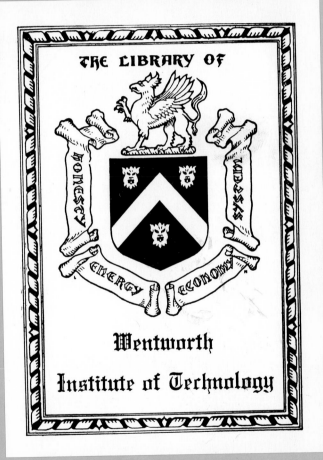

The MAN *with* NO
ENDORPHINS

The MAN *with* NO

ENDORPHINS

and other reflections

on science by

JAMES GORMAN

Viking

VIKING

Published by the Penguin Group
Viking Penguin Inc., 40 West 23rd Street,
New York, New York 10010, U.S.A.
Penguin Books Ltd, 27 Wrights Lane,
London W8 5TZ, England
Penguin Books Australia Ltd, Ringwood,
Victoria, Australia
Penguin Books Canada Ltd, 2801 John Street,
Markham, Ontario, Canada L3R 1B4
Penguin Books (N.Z.) Ltd, 182–190 Wairau Road,
Auckland 10, New Zealand

Penguin Books Ltd, Registered Offices:
Harmondsworth, Middlesex, England

First published in 1988 by Viking Penguin Inc.
Published simultaneously in Canada

All the essays in this book first appeared, some under different titles, in
Discover Magazine.

Grateful acknowledgment is made for permission to use the following
copyrighted works:
    Excerpt from "Flushed From the Bathroom of Your Heart" by Jack
Clement. © 1967 by Jack Music, Inc.
    Excerpt from "You Don't Mess Around with Jim" by Jim Croce. © 1971,
1972 DenJac Music Co. & MCA Music Inc. All worldwide rights administered
by DenJac Music Co. Used by permission. All rights reserved.
    Excerpt from "Drinkin' Wine Spo-Dee-o-Dee, Drinkin' Wine" by Granville
"Stick" McGhee and Mayo Williams. © Copyright 1949, 1973 by MCA Music
Publishing, a division of MCA Inc., New York, NY 10019. Copyright renewed.
Used by permission. All rights reserved.
    Illustrations by Joe Le Monnier from *Discover* Magazine, May 1986.
© *Discover* Magazine 1986.

LIBRARY OF CONGRESS CATALOGING IN PUBLICATION DATA
Gorman, James, 1949–
    The man with no endorphins.
    1. Science—Anecdotes, facetiae, satire, etc.
I. Title.
Q167.G67   1988   500   87-40300
ISBN 0-670-81842-9

Printed in the United States of America by
Arcata Graphics, Fairfield, Pennsylvania
Set in Century Schoolbook

For my wife

Contents

Introduction

A number of years ago I wrote a short news item about something called a bezoar stone. At the time I was a serious science writer, which is to say that I wrote articles about serious science and I was serious while I wrote them. I had tackled some big subjects—stars, galaxies, black holes. But I was beginning to tire. I was beginning to feel that if you'd written one article about spiral galaxies, you'd written them all. Consequently, bezoar stones caught my attention.

They're remarkable objects. To quote myself in *The Sciences* (May/June 1979), "Bezoar stones are not actually stones at all, but concretions of partly digested and calcified hair that bump and jog for years along the alimentary canals of antelopes, goats and other ruminants." In other words—hair balls. They were renowned, somewhat renowned anyway, because legend had it that they offered protection against poisoning. Potentates of one sort or another used to dip these stones into glasses of wine they suspected might have been tampered with. Then they had somebody else drink the wine.

I wrote a news story about the stones because scientists had discovered that bezoar stones really worked, at least for one poison. They really did detoxify arsenic, somehow absorbing or trapping toxic molecules. The secret ingredients were partly digested hair and a mineral called brushite. All of which made for a pretty good story with a bit too much of this kind of talk: "The disulfide linkages in the stone's calcified hair protein are broken down, exposing sites for binding between sulfur and arsenite." (You can see why I gave up straight science writing.) However, what was most interesting to me about bezoar stones never got into the article. How, I wondered, did anybody get the idea to use bezoar stones in the first place?

To a man faced, day in and day out, with the task of writing about the nature of space-time, here at last was a truly interesting question. And I wasn't the only one who thought so. Speculation about the first man or woman to find and use a bezoar stone caused a great deal of talk in the magazine office. After all, one does not just pluck a hair ball from the stomach of a goat and plop it in the king's wine. One has to find it, figure out that it's good for something, and then figure out just what that something is.

I discussed with a friend of mine how the discovery might have been made. We understood how folk medicine worked with plants. People were eating plants anyway. They noticed that some of them tasted good, some of them made them sick, and some of them made them want to lie around all day listening to the Doors through headphones. It was obviously very important to remember which was which. But bezoar stones were more problematic. True, people were already eating goats, but whether a bezoar stone is a part of a goat or just *in* a goat is a matter for debate.

Consider the problems of going from the raw material to the fully evolved anti-poison amulet: Suppose the first one was found when somebody at the annual goat-slaughter festival pulled it out of his *tripes au chevrier*. Suppose further that the thing was assumed to have magical or medical powers because it was an anomaly, not explicable by the known laws of digestion. Suppose finally that the goatherds figured that if the gods had put these stones in goats' stomachs, they'd done it for a reason, probably to save the king's life.

Still, somebody had to connect the stone to poisoned wine, and test it. That's where Karl Popper comes in. Karl Popper is a preeminent philosopher of science and it's my experience that he always comes in somewhere if you let him. He and other historians and philosophers of science have discredited the old notion that scientists (and by extension goatherds) proceed by observing and cataloguing facts in an objective fashion until some conclusion about the workings of nature emerges. They've shown that researchers approach their subjects guided by goals, by theory and hypothesis, by available funding.

Funding. Funding is the key to bezoar stones. I didn't recognize it at the time but I've been thinking about bezoar stones now, off and on, for almost ten years. And what I've concluded is that then, as now, research energy was directed to the ailments of people who could afford to pay for a cure. Witness the modern search for a Type A personality antidote. You don't hear a lot about people in the Third World dying of Type A personalities. You know how many calories per day it takes to have a Type A personality? Similarly, in the old days, nobody was slipping arsenic into goatherds' wineskins. It was the potentates who were suffering the poisoning pandemic. This brought a constant stream of goatherds (and everyone else) to the palaces, hawking various herbs, powders, and goat parts as antipoison charms, and hoping for gold. One of them got lucky. Others were not so fortunate. I have to believe that peer review in those days was tougher than it is now, that when a man walked into a potentate's court and said, "Your Highness, my amulet will protect you from poison," the standard response was not to send his paper out for comment, but to chain him to the table, give him a cup of poisoned wine, and ask for a demonstration.

It was this kind of thinking that enabled me to solve the bezoar stone problem—maybe. But it also led me astray. Once you start spending your time with potentates and amulet salesmen it's hard to go back to disulfide bonds. You get an inkling of life beyond spiral galaxies (I say this metaphorically; in actuality I've never had any contact with beings outside of our own particular spiral galaxy), and you want to see more. It was just an inkling, of course. I had no idea at first where the fascination with what we might call hairball science would lead me. It (the inkling) needed a few more years of serious science writing and then a few more years of less serious science writing finally to accrete, like hair in a ruminant's stomach, into the collection you hold in your hands. But accrete it did. And now here it is, twenty-eight pieces on science, and not a spiral galaxy in the bunch.

I can't take full credit for the book, nor would I want to. As with any undertaking of this magnitude, there are a number of people other than myself who deserve thanks and recognition.* My parents not only brought me into the world, they sent me to college, in the hope that one day I would do something worthwhile. Perhaps one day I will. After them, there's Gil Rogin. By jumping to him I have, it's true, skipped over the many teachers who influenced my writing, but none of them actually paid me. Several years ago Gil Rogin took a chance and asked me to write, for money, a column in *Discover,* the magazine in which these pieces appeared from 1985 to 1987 (some in slightly different form than they are here). He is, I suspect, the only editor who would send a writer to southern California to

---

*Will Cuppy, for instance. I adopted the habit of reckless footnoting in shameless imitation of Cuppy's funny pieces on scientific subjects.

interview the manufacturers of toilet-tank fill valves. Of course, I'm probably the only writer who would go. Thank you, Gil.

Thanks also to Marilyn Minden, without doubt the world's best copy editor, who protected and improved these pieces, to Arnold Roth, who illustrated them in the magazine, and did the cover of this book, and to everybody in the *Discover* copy room, who, when I was paid by the word, put squeezes on my columns to get the highest possible word count for the space I was allotted. Tom Dworetzky (and on occasion Sally Dorst and Allan Chen) checked the few facts I was forced to include in my columns, thus keeping me out of court (so far). My agent, Kris Dahl, figured out how to sell the pieces yet a second time, and my editor at Viking, Dawn Seferian, had the enormous good taste to buy them and put them in book form. She has promised me that as a result of this collection I will become more famous than I am now. I have no doubt that she's right.

Following tradition, I have left the best until last—my two most important editors, my wife, Kate, and my friend and fellow writer Richard Liebmann-Smith. Both heard or read most of these pieces before they were finished, and both offered valuable editorial (and other) advice. Richard has helped me with my writing for many years, sometimes to the point of giving me jokes and titles. It was with him that I first discussed the whole bezoar stone problem. Over all this time his critical judgment has been, with one egregious exception, unerring, and his generosity with his time quite amazing. Since I have never responded well to criticism of any sort, both Kate and Richard may, at times, have viewed their task as a thankless one. I have now proved them wrong.

Of course, with all this help, all these people who've had some part in putting me where I am today, there is still, as every writer (and every reader of introductions) knows, somebody who has sole and final responsibility for what goes down on paper . . . and it's not me. No sir. If I have to share responsibility for the good stuff, I'm not going to eat the errors. The funny parts are mine, all right, but if there's anything wrong with any of these pieces it's Richard Liebmann-Smith's fault.

James Gorman
May 1987

The MAN *with* NO
ENDORPHINS

In Praise of the Pump

A momentous event—well, an event, anyway—has occurred in the world of science—well, of toothpaste, to be precise. The dentifrice industry, long represented in most homes by tubes of toothpaste squeezed in infuriatingly incorrect fashion by other members of the household, has now come up with its version of the better mousetrap: the toothpaste pump.

Actually there are several versions, from Check-Up, Colgate, Crest, and Aim. In the past year the companies that make these brands (respectively, Minnetonka, Colgate-Palmolive, Procter & Gamble, and Lever Brothers) have brought out sleek-looking toothpaste pumps, shaped like small, blunt-tipped rockets, as an alternative to the old-fashioned tube. All except Procter & Gamble (which early in the year was still testing the Crest pump in Maine, flying, so to speak, in the teeth of New England conservatism) have their product on drugstore shelves. And, by the companies' accounts, all their pumps are doing very well.

This isn't the sort of news that can be reported objectively, the way one might report the fact that a bubble memory for computers has been perfected, or that someone has discovered that the universe is twenty billion years old. Those are neutral stories. They may be interesting, or even important, but except for the man who has the patent on the bubble memory, or the woman who will get the Nobel Prize for figuring out the age of the universe, people who wake up the morning after the news hits the front page won't find their world changed.

But when someone wakes up and finds, not a tube, lying crumpled, encrusted, and topless on the lip of his sink, reminding him of his slovenly housekeeping and, if he is an introspective person, of his own mortality, but a pump—vertical, cylindrical, clean, upright, unsullied by leaking dentifrice, making him feel, if not immortal, then at least like someone who has a good handle on both dental and domestic hygiene—his life may be permanently changed.

You can see my bias. I'm admittedly pro-pump, partly because I'm a sloppy tube-squeezer. Neat tube-squeezers are generally anti-pump, since, in my opinion, they're already obsessively squeaky-clean in their habits and don't need technological help. But there are other, more objective reasons to praise the pump. It represents the best kind of invention. It's non-electronic. It doesn't have to be programmed. It's not turbo-charged. It is, like the nutcracker and the can opener, a "low-tech" invention.

I'm aware that "low-tech" isn't a common scientific term, like entropy or lymphocyte, so I'll try to define it by example. These things aren't low-tech: the space shuttle, cellular phones, and personal pulse monitors. These things are: steel wool, scissors, ball-point pens, staplers, and the Veg-O-Matic. More generally, a low-tech invention makes no noise and requires no maintenance. It's undemanding, and it usually fits in a drawer or a medicine cabinet.

I'm not altogether immune to the charms of high-tech consumer goods. I own a computer. There are times when I have lusted in my heart after compact disc players and modular televisions, after a home entertainment center so lavish and complex that it would require not an instruction manual but a pilot. But the truth is that my electronic lust has never approached the intensity of the desire ignited in me by the kind of low-tech wonders advertised on late-night television along with albums of old songs, those magical items that are, mysteriously, "not available in stores"—lint brushes and miracle kits to repair cigarette holes, non-stick frying-pans, choppers, dicers, graters, and peelers. There is a Newtonian, or perhaps even pre-Newtonian, innocence to a well-designed carrot peeler. Poems could be written about good spatulas. Garlic presses, pizza cutters, and, perhaps most of all, those little round-ended, serrated, grapefruit tools should be preserved, if they aren't already, in the Smithsonian Institution. I would argue that there is no greater tribute to the benevolent power of the human imagination than a store full of kitchen utensils.

The spirit of low-tech isn't limited to the kitchen. There are wrenches, ratchets, and ball-peen hammers for people who like that sort of thing. And I discovered a new breed of sock recently, made with a mixture of wool and polypropylene, that actually keeps my toes warm in cold weather. I'm far more grateful for this contribution to human life than I am for my word processor. With warm toes, I could write happily with a quill pen.

Most of these wonderful inventions are commercial in nature, the product of the profit motive and not of a humanitarian concern for my toes or anybody else's. So also with the pump. Its origin lies in a major problem faced by toothpaste science, otherwise known as the billion-dollar-a-year dentifrice industry. That problem is success. Ninety-eight percent of the American people brush their teeth. While it's true that a mere 95 percent do it with toothpaste (the others must use baking soda, or something), that still doesn't leave much room for expansion.

So, in the past, the industry, and various inventors inspired by the importance of tooth-brushing to daily life and the possibility of getting rich, have tried improvements, which, if they had been of any use, might also have counted as low-tech gifts to humanity. They include aerosol cans for toothpaste (they didn't succeed—no doubt because aerosols are, if not high-, at least middle-tech); clear, green, bright red, grey, blue, white, and striped dentifrices; mechanical Tube-Wringers (to get the last squeeze); brushes that come with their own paste in them; a range of ingredients, including fennel, limestone, vitamins A, D, and C, clay, and sea salt; and a list of chemicals too long, confusing, and alarming to report. My favorite among the exotic toothpastes, which I haven't yet tried, but which one unbiased dentifrice reviewer reported "does leave the teeth squeaky clean and the breath gaspingly fresh," is Monkey Brand, from India. It isn't a paste but a black—yes, black—powder.

The pump is a different matter. It's an invention so obvious, so simple, and so potentially lucrative that it was invented by West Germans—or at least some of the pump designs were. In fact, pumps have been common for years in Europe, which clearly leads the U.S. in the dentifrice race. Now that the U.S. has jumped into the competition, American manufacturers are reluctant to discuss the design of their pumps, presumably to keep them out of the hands of the Russians. However, in a free society, toothpaste pump designs, like plans for atomic bombs, can often be retrieved from public documents. After considerable effort, I managed to obtain patent reports on some of the pumps.

There are two general kinds, as near as I can figure it. One uses vacuums and valves (Check-Up and Crest), the other is mechanical (Colgate and Aim). Colgate's pump, for example, is a slightly improved version of a caulking gun. Each time you press on the trigger, a piston moves a short distance up a pole, pushing ahead of it the toothpaste, which emerges onto your brush. The Crest pump, a vacuum-valve type, seems more complex, like an artificial heart operated by finger power. It has an upper chamber, from which the paste emerges onto the brush, and valves to let paste in and out of the chamber. When you push the button, the valve leading back into the body of the pump closes and the one leading to the spout opens. The force that you exert reduces the volume of the chambers, thus squishing out the decay-preventing dentifrice. When you take your finger off the button, the volume increases, the valves change position, and new paste is sucked up into the chamber, ready for the teeth of whoever shares your bathroom.

The pumps work. I have personally tested three of them (Procter & Gamble couldn't find an extra Crest pump to send me, and I couldn't make it to Maine), and I can say that those three all get the toothpaste out. Nothing is perfect, of course. I think that using vacuums and valves is already sneaking up into middle-tech, and Check-Up and Aim have flexible plastic tops that will squish, more or less, if you push on them. (Squishiness reminds me of tubes. It lessens the clean, rigid feeling of good housekeeping you get from the Colgate pump.) The Aim pump also has a little tab that you're supposed to stick back into the spout after each use, which is not in the spirit of low-tech. What I myself found most unsettling in my research is that the top of the Check-Up pump bends way back as you push out the toothpaste. To the consumer, at least this consumer, using that pump feels a bit like breaking the neck of a small animal, which is not the way I like to start the morning.

As a personal choice, I have to go with Colgate, with one reservation. Their instructions show a finger being used to work the pump. They're wrong. The only sensible way to work the pump (I have corroborating opinions on this from other consumers) is with the thumb. Why Colgate couldn't figure this out is beyond me.

Of course, the thumb/finger question is nothing compared to the wars that erupt over how to squeeze a tube. I still say that as inventions the pumps are right up there with the garlic press and the nutcracker. And since their inventors are never going to get Nobel prizes or tenured faculty positions, I would like to provide them with a small measure of fame to go along with the piles of money they already have. In gratitude for cleaning up my sink, and in small recognition of his work for the good of the dentifrice-using portion of humanity, I would like to thank, right here in print, the inventor of the Colgate pump (I can't mention everybody), Alfred von Schuckmann of West Germany. With toothbrush in one hand and "dispenser for, in particular, pasty substances" (U.S. Patent No. 4,437,591) in the other, I salute you.

The DNA of the DAR

I for one am very proud of the Daughters of the American Revolution. That hasn't always been the case. I used to think they were stuffy, and altogether too absorbed in their genealogies, or, looking at it biologically, their genes. I didn't mind their looking after the future of those genes, getting out there in the ecosystem of garden clubs and tea parties and scrapping for the kind of biological and social advantage that would enhance their descendants' chances of survival. It was the excessive looking backward, to their ancestors, that seemed to me biologically useless.

I was wrong. The Daughters may well be stuffy, but their family trees certainly aren't useless. The DAR has brought this point home in its latest project, neither a cookbook nor a monument but a genetic study. In cooperation with the Vanderbilt medical school, the Daughters are using their genealogies to trace, not blue blood, but the inheritance of disease. Questionnaires have been sent out to members as part of the DAR Family Tree Genetics Project, asking them to provide a genealogy covering at least three generations that specifies each person's medical problems, including known genetic disorders as well as other diseases. What the scientists are looking at is, in an indirect way, the DNA of the DAR.

When the project was announced, the president general of the DAR, Sarah King of Murfreesboro, Tennessee, pointed out that it could show patterns of inheritance of disease that would be valuable to the medical profession and the public at large. Taking the high ground, she told the membership, "We foresee this program as a boon, not only for our own families but to mankind as a whole. We are fortunate in that we know from whence we came. Our research will serve as an inspiration to others."

I like a president general who knows how to talk like a president general. Furthermore, I like an organization that doesn't indulge in false modesty. And the Daughters are rightfully proud of themselves for getting involved in a project so important that it makes them not only daughters of revolutionaries but revolutionaries themselves. It suggests what could be a tremendously fruitful connection between the world of science and the world of clubs. There is the potential here to liberate population geneticists from their interminable and tedious involvement with fruit flies, to let them say good-bye to *Drosophila* and hello to the DAR, the Boy Scouts, the National Basketball Association, and the Académie Française.

I hope this is what Mrs. King had in mind when she talked about inspiring others, because that's what comes to my mind—a whole new field of science. Call it organizational genetics, or, better, club genetics. Think of it: instead of laboring over a dissecting microscope, checking flies for plum-colored eyes and curly wings, a researcher could study the inheritance of height in professional basketball players. Of course there would be some problems to overcome. The questionnaires would have to have instructions explaining that the traditional qualitative terms used by fans and sportscasters to describe athletes—tall, really tall, and too tall (as in Ed "Too Tall" Jones, who played pro football, in complete disregard for his nickname)—aren't adequate for science. Researchers need feet and inches, if not centimeters. But the basketball study, it seems to me, would be what a scientist or a betting man might call elegant, particularly since it has the advantage of making season tickets an expense that could legitimately be included in a grant application. And for anyone who has sat through the endless quantitative mumbling of a scientific meeting, the project offers the possibility that, during a presentation at the annual meeting of the American Association for the Advancement of Club Genetics, one of the new breed of researchers, instead of asking for an inscrutable slide of some poor *D. melanogaster*'s chromosomes to illustrate his point, will call out from the podium, "Let's go to the videotape . . ."

One could study not only ancestors but also descendants, doing with the DAR or the NBA what Gregor Mendel did with his peas. What do you get, for instance, when you cross the offspring of a center with the offspring of a guard? A forward? And what about real hybrids? What would be the result of a DAR/NBA cross? Perhaps in the first, or F1, generation, there would be a mix—members of the NRA (National Rifle Association) and the ABA (American Bar Association)—while in the F2 generation the offspring would revert to type, DAR and NBA members wondering how in the world they were born into families devoted to guns and the law.

These are only the most obvious projects. Once the discipline of club genetics begins to flower, researchers will be able to tackle the stickier challenges posed by sociobiologists, who have, at one time or another, suggested that almost any human trait, from sexual preference to religious belief, is written in the genes.* My favorite among the qual-

---

*Much to my disappointment, nobody has suggested that there's a separate gene for each religion, though I suspect that's the case.

ities the sociobiologists have described is altruism. Altruism, of course, is selfless behavior, of the sort the DAR has displayed in conducting its study.

In his book *On Human Nature,* E. O. Wilson of Harvard described two kinds of human altruism, hard-core and soft-core. The first involves sacrificing yourself for a close relative who doesn't pester you for money all the time, and it brings no reward other than the survival of someone who shares your genes. This is, in fact, altruistic. Soft-core altruism, on the other hand, brings some kind of non-genetic benefit, like a medal or reward money or a magazine article praising your organization for its public-spiritedness, and is, at base, selfish. (This constitutes a cynical view of human goodness to which neither I nor, I'm sure, the DAR subscribe.) One of the great benefits of club genetics would be that it would offer researchers the opportunity to do *the* altruism study—a survey of the one organization that more than any other purports to be devoted to good, unselfish acts. How could any scientist resist the chance to publish a paper with the title "The Sociobiology of Altruism in the Boy Scouts of America"?

Americans for Common Sense would be another nice group to study, although I would prefer an organization called Americans *with* Common Sense. I'm sure there's something to be done on WITCH (Women's International Terrorist Conspiracy from Hell), if only to find out whether its members have a genetically determined talent for creating great acronyms. A comparative geneticist might discover what separates the Elks from the Lions. And I would be very grateful if someone would take a close look at the Académie Française. This is the organization that defends French culture (including the famous "choking r," which sets apart the true French accent from all imitations) and admits to its ranks the major French intellectuals, like Claude Lévi-Strauss. This is what I want to know: Is the ability—or for that matter, the desire—to speak French with your lips puffed out and your eyes half closed inherited? And, as a subsidiary but related question, Is there a gene for pomp?

Eventually, thanks to the avant-garde action of the DAR, we should be able to do scientifically what the boards of exclusive clubs have been doing all along on an ad hoc basis—decide what a person is like, genetically and personally, on the basis of his or her club memberships. For instance, a person who belonged to both the DAR and the Académie Française could safely be presumed to be a descendant of General Lafayette and to have two dominant pomp alleles.

I suppose that by then anyone putting comments like these down on paper will have to disclose what clubs and organizations he belongs to. To be fair, and in the spirit of full disclosure, which I support for everyone else, I'll do this now. As may already be obvious, I am not, nor could I be, even if I were a woman, a member of the DAR. I was never a Boy Scout. The groups I do belong to are: the National Association of Science Writers, the Authors Guild, and Trout Unlimited, which, I would like to point out, is an organization for, not of, fish.

Guess What's Coming to Dinner?

I just knew it wasn't wrong to eat meat. I had what you might call a gut conviction. It came over me every time I was in close proximity to rare steak or roast pork rubbed with sage, the fat crinkled and brown from the oven. Braised short ribs with onions acted even more powerfully to make me see the virtue in carnivory. And in the presence of a butterflied leg of lamb riddled with garlic slivers, grilled over charcoal, I began to discover countless reasons why my particular arteries would be spared from the atherosclerotic ravages of good food.

To a skeptic, or a cardiologist, I must sound like a man tempted by adultery who develops a belief in open marriage. Science and medicine have spoken on red meat with one voice, and in much the same tone that Sister Miriam used to talk to the sixth grade at St. Justin's about kissing. It's bad for you. It causes heart attacks, triple bypasses, and the indignity of having to put on jogging shorts and have high school girls laugh at you. And yet a good steak, like a kiss, is hard to turn down. I don't know if the Catholic Church has revised its teachings on puppy love—at a certain point in my life I became more concerned with heart disease than the fine points of necking—but I do know that heretical voices have been raised on the subject of meat in the *New England Journal of Medicine,* which, if not a medical bible, is at least a catechism.

The voices are those of a physician, S. Boyd Eaton (pronounced eatin') from Emory University in Atlanta, and an anthropologist, Melvin Konner, author of *The Tangled Wing* (which isn't about chicken parts). They asked this compelling question: What should we have for dinner? And the answer, medically speaking, was meat. Examining the diets and ways of life of prehistoric hunter-gatherers, as well as modern ones, like the !Kung bushmen, Eaton and Konner concluded that our species adapted over thousands of years of evolution to a diet that includes a lot of meat, fruit, and vegetables, and no grains or dairy products, to say nothing of Pouilly-Fuissé. Modern hunter-gatherers who eat this way don't suffer the ailments of television producers, like heart disease, high blood pressure, diabetes, and some kinds of cancer. (Of course they aren't television producers.) And Cro-Magnon man, the father of big-game hunting, although he had no table manners (actually, he didn't even have tables), was apparently also healthy.

Now, since our bodies are essentially the same as those of our ancestors, one can only conclude, with a leap of joy, that it's time for a guilt-free porterhouse. However (there's a however in every medical paper just when you get out the steak knives), the meat hunter-gatherers eat, and ate, comes from what the authors call "free living" animals, animals with antlers and fleet feet, and also low levels of fat (particularly saturated fats) and high amounts of a chemical suspected to combat atherosclerosis—perhaps on the basis of its medical-sounding name, eicosapentaenoic acid. This kind of meat seems to be just as good for you as bulgur in tofu sauce. Presumably, animals like gazelles that are always getting scared out of their wits and having to make those remarkable leaps when a lion or David Attenborough pokes his head out of the tall grass just never get the chance to put on much harmful fat. Fatted cattle, on the other hand, live in fear-free abundance, designed, of course, to fat them. The result is that they give us heart attacks.

There are two possible solutions to this problem. One is to hire a lot of people to go to Kansas City in lion masks and scare beef. The other is more realistic, but not much more. Truly health-conscious people will just have to stop doing exegeses of food labels and learn to hunt. If there's any message to be twisted out of Eaton and Konner's thoroughly sensible paper by unscrupulous exaggeration and oversimplification, it's this: you can have your pork and eat it too, as long as it comes from a wild boar. To me, with my weakness for all parts of the pig (the bacon part, the chop part, the sausage part), the thought is a liberating one, but I'm sure it will upset those more serious about their health. *Vegetarian Times* may not know it yet, but with the publication of Eaton and Konner's paper, the leading edge of the nutritional revolution has shifted from fruitarian extremists to the good ol' boys with gun racks in their pickups.

By rights, venison should take over from Tofutti, hunting camps should be the new health spas, and the thin and the chic should rediscover *cuisine sauvage*. The news should also affect the other segments of the health industry. For example, bird hunting might well replace aerobics—providing both exercise and a healthy meal. But I have my doubts. What would become of Richard Simmons? I don't see him, or his devotees, toting shotguns through soggy thickets after woodcock. And I don't believe Jane Fonda will tote a 30.06 into the wild to bag supper. In fact, most nice people, even if they do eat meat, prefer to have someone else do it in.

Well, there's an alternative, which allows one to experience the benefits of the Free-Living Diet with no moral qualms, but I hesitate to mention it. Actually, that's not true, I've been waiting for a chance to mention it for seven years, ever since I read, in the Spring 1978 issue of the *Co-Evolution Quarterly,* what I can say, without fear of contradiction, is the single weirdest piece of food writing ever to be published in a developed country. The title of the article was "How to Use Road Kills."*

This isn't a joke. The article was completely serious. It included the usual tips on skinning, reviews of past repasts (including a hellbender—a huge, ugly, aquatic salamander), a poem by Gary Snyder about eating road kills (and a good poem at that), and this remarkable observation: "Perhaps you are new to the delights of carrion eating and . . . a little unsure of your judgment in these matters. Just what are the consequences of eating spoiled meat? Apparently there are none if it is sterilized by cooking. According to my research, 'spoilage' is a relative, cultural term."

This may be taking cultural relativism too far, and in the wrong direction, but the drift of the article is sound. Our pre-Cro-Magnon ancestors may have been not hunters but scavengers. They were availing themselves of creatures that had been run over by saber-toothed tigers, not Chevys, but the principle is the same. There's a lot of high-quality, low-fat protein out there on the shoulders of I-80 and somebody in some phylum or other is going to eat it, so it may as well be you. It's certainly not going to be me.

No, although I do care about cutting down on fats, and although modern scavenging has logic and ethics on its side,

*As I discovered later, eating road kills is commoner than you might imagine. The writer John McPhee once consumed his portion of a roasted road-killed weasel. He recorded the meal in a 1975 *New Yorker* profile of an unconventional Georgia zoologist named Carol Ruckdeschel, who said that she hadn't bought meat for a year, "except for some tongue."

and may, in fact, be the ultimate response to the throwaway society, I'm afraid I have neither the moral fiber nor the stomach for it. I disappointed Sister Miriam many times, although not as many as I would've liked, and I've never been able to stop eating anything I liked. Besides, I have a long, unrequited desire to own a pick-up truck. Eaton and Konner have set me free. I'm going to resume the Cro-Magnon life, with the aid of four-wheel drive. And when one of my Volvo-driving, leaf-eating friends sees me stopped at a light in my new vehicle, nursing a beer, and raises an upwardly mobile eyebrow, I'll just tip back my John Deere cap, have another swallow, and explain to him, in terms that no health-conscious yuppie could question, that it's O.K.—it's all part of my new diet.

The Implausible Dream

Pharaoh (on couch): *In the dream I'm in this field, and there are all these kine.*
Joseph: *Kine?*
Pharaoh: *That's right, kine, seven fat ones and seven thin ones. And then the thin ones eat the fat ones. Kine—eating each other. Why would a person dream such a thing?*
Joseph: *What are kine?**

O f course, everyone knows that Joseph wasn't a Freudian analyst. He practiced in a different tradition, one in which people were familiar with kine and what they represented. Joseph didn't look for repressed sexual urges in dreams, because—my authority here is Norman Mailer—repression of urges wasn't in vogue in ancient Egypt. Still, as interpreters of dreams, Joseph and Freud shared the belief that, whether your dreams are telling you that you're a latent foot fetishist or that you should store up a lot of grain for seven years of famine (the thin kine), they're telling you something.

Recently, however, some scientists who study the brain,

*Scientists have since learned that kine are cattle.

17

where it's now universally agreed that dreams originate, have suggested that dreams are a kind of random mental noise, or perhaps neurological junk the brain is discarding. This is going too far. I don't think there's any question that we could all do with fewer phallic symbols in our dreams, but in these new ideas a cigar isn't even a cigar, just an odd impulse leaping a synapse while we sleep.

Among the leading proponents of this sort of thinking are psychiatrists Allan Hobson and Robert McCarley of Harvard Medical School, who concluded, on the basis of sleep research, that there's a dream generator in the brain stem that causes more or less random firings of neurons. The forebrain—which if not the mind is at least more like it than the brain stem—is then forced to interpret these incoherent signals. This is why dreams are filled with people without pants on, relatives who are giant lobsters, and constantly shifting scenes. The forebrain is in the position of a director who must stage nonsense and make the audience think it's seeing a play, a circumstance Hobson and McCarley argue is as common in neurobiology as it is in the theater.

Another non-Freudian idea has been put forth by two biologists, Francis Crick, the Nobel laureate, and his colleague Graeme Mitchison. They base their hypothesis partly on Hobson and McCarley's work, but say that what's going on during the stimulation of neurons is a "reverse learning" or "unlearning" of stuff that the brain needs to lose to keep running clean. Their notion is that during sleep the brain stem sends out stimuli that stir up the neuronal dust, which the forebrain stuffs into dreams and tosses out. Thus, dreams aren't the royal road to the unconscious but the trash bags of the brain.

If true, this is very disappointing news. But perhaps it isn't true. The theories do seem to run counter to the past experience not only of prophets, but of poets, scientists, and, to take an example close to home, me. Why, if my dreams consist of random noise, are the old girl friends who appear in them my own, and not someone else's? And what conceivable reason could there be for my having spent the past twenty years of my life, off and on, confronting the awful fact (while dreaming) that although I'm on a stage and have just received my cue, I've no idea what play I'm in? Crick says that recurring dreams are those that wake the dreamer and, so, cause him to learn instead of unlearn them. A recurring dream is thus a kind of neurological flypaper, and the brain is compelled to indulge in an endless slapstick routine trying to get it off its fingers.

Maybe. But that doesn't solve the larger and far more interesting problem of Samuel Taylor Coleridge, who claimed to have dreamed the poem "Kubla Khan." I know Coleridge had opium to help him, and I'm sure that does something weird to the neurons. In fact, as I understand it, that's the whole point of opium. Still, Coleridge did have this incredibly rich dream, with not only visual images but accompanying verse:

> *In Xanadu did Kubla Khan*
> *A stately pleasure dome decree:*
> *Where Alph, the sacred river, ran*
> *Through caverns measureless to man*
> *Down to a sunless sea.*

He woke up and wrote down this much, and a bit more, including some gardens, woods and dales, caves of ice, and dancing rocks. Tradition has it that he would have got the whole dream down had he not been interrupted by a stranger from Porlock.

Depending on your taste in poetry you may be disappointed or delighted that Coleridge was interrupted, but whatever your opinion, the thing does have meter, it rhymes, it hardly seems the work of neurons firing away like the proverbial monkeys tapping on typewriters. I suppose you could argue that Coleridge, being a poet, had a highly skilled forebrain that was able to ghostwrite the dream on the basis of scattered neurological impulses, a pleasure dome here, a sacred river there. I might believe this for "A damsel with a dulcimer/ In a vision once I saw," which seems like standard poetic boilerplate, but what about: "But O, that deep romantic chasm which slanted/ Down the green hill athwart a cedarn cover!" You just can't get "athwart a cedarn cover" from random neuron firings, no matter how hard you try.

"Kubla Khan" is a tough one for Hobson and McCarley, but it doesn't necessarily refute the clean-up scheme of Crick and Mitchison, who could claim that this is a poem that was supposed to be forgotten. Let's suppose for the sake of argument that they're right. The brain is like someone in an office—all day long it works, learning things, writing things if it's a certain kind of brain, perhaps doing experiments in organic chemistry if it's another kind. During the course of its work, it accumulates the equivalent of scrap paper, little things with jottings on them that say "Column due in two days" or "Structure of benzene? Cube?" Then after all this hard work, while the day part of the brain is enjoying well-deserved sleep, the night crew cleans up. Here's the problem: How can you trust it to know what to throw out?

Take the case of Friedrich Kekulé, the famous nineteenth-century organic chemist from Germany. For years Kekulé attempted to find the structure of the benzene molecule. Always, he tried to stay awake, for fear of what he might lose in a dream. Then one day, while thinking about benzene—he was always thinking about benzene—he fell asleep and had a dream in which a snake was eating its own tail, forming a ring. In Crick and Mitchison's terms, while the part of Kekulé's brain that actually did organic chemistry was dozing, the cleaning team was rummaging around looking for stuff to throw out. "Snake?" it said. "What's this snake doing here? We're doing benzene, right? Chuck it." Fortunately, Kekulé woke up, snatched the dream out of the trash, and realized that the carbon atoms in benzene could form a similar ring, which, in fact, they do.

That's one way to tell the story. The traditional interpretation of this event, which I hold to, is that Kekulé's brain wasn't cleaning itself out but being creative, and that it meant for him to use the snake dreams to make a name for himself in chemistry. I'm not, of course, a Nobel laureate. I'm not even a laureate. And Hobson and McCarley are both doctors. Furthermore, everything I know about neurobiology you have just read. So it would probably be foolish of me to come right out and say these guys are wrong. Instead, let me just say that if *I'm* wrong and brains *are* throwing away these kinds of ideas, maybe we should find a way to stop them.

The Age of Aquariums

T here comes a time in everyone's life to keep fish. Just as there comes a time, later, to sell the aquarium at a garage sale. The second event is always an occasion for celebration, but each person reacts to the first in his own way. The chuck-and-chance-it aquarist* meets his fate lightly—buying a fish, chucking it into the aquarium, and chancing it. Of course, in reality, it's the fish who chances it, which is why these aquarists don't lose sleep over the pH of their water, its hardness or softness, or the fact that *Betta splendens* (known in our house as Danny) has a taste for live tubifex worms.

Then there are those of us who take fish, and life, more seriously, those of us to whom an aquarium brings a burden of ecological responsibility similar to that felt by the head of the Audubon Society, or God. In our five-, ten-, or twenty-gallon tanks there are—in addition to the fish—plants, bacterial diseases, and enough protozoan parasites to support an entire university biology department. The average com-

---

*I've borrowed this phrase from people who don't keep fish, but go fishing. A "chuck and chance it" angler chooses lures with cavalier disregard for their appropriateness. This is simply a different form of disrespect for fish.

munity tank cedes nothing to "Dynasty" or "Dallas" in intrigue and conflict. Different fish seek different habitats, territorial battles occur, and, just as on TV, the big ones eat the little ones. An aquarium offers lessons in ecology, biology, chemistry, and personal space.

I came to fish late in life, when I had an income to support them. I purchased, ostensibly for my daughter, a Siamese fighting fish (the aforementioned Danny, since buried in the yard to incantations of "Bye-bye, Danny, bye-bye, fish"), a bowl, and a book. Bowl, book, and fish aren't the ingredients of some piscine excommunication, but the opposite, baptism. For those of you who don't yet own fish, the book is the purchase to avoid. Mine explained to me that in an unheated bowl my betta would grow sluggish and starve. I needed a heater, and since heaters aren't made for bowls and will stratify uncirculated water, I needed an aquarium and a filtration system, which had to be gentle because bettas like calm water. Of course I also needed plants and an aquarium light. I bought it all. Had I remained unlettered about keeping fish, I would have been satisfied with the bowl, and when the fish died I would have just assumed it was his time to go. Then I would have been able to put dried flowers, or rocks, or little plaster figurines of cocker spaniels in the bowl, and to resume life. With fish, ignorance may not be bliss, but it's cheaper than knowledge.

As happens, one fish led to another, and the aquarium was struck by death and disease; I dosed my fish with tetracycline and malachite green to combat bacterial fin rot and ich (*Ichthyophthirius multifiliis,* a parasitic protozoan). I bought an 800-page compendium on fish and their diseases. And I subscribed to *Tropical Fish Hobbyist,* which revealed to me the lengths to which fish obsession could be taken. The cover of the April issue consisted of eight photographs of different platies (fish). And the explanation of the cover began with this sentence: "What can one say about *Xiphophorus maculatus,* the platy or moon, that hasn't been said a thousand times over . . . ?"

What indeed? And where does one end, once one starts to think, and write, about platies in a certain way? There's the old saying about greener grass, and it's also true, at least in the Free World (this is in fact the foundation of both the Free World and the tropical fish business), that no matter what one has, one always wants more. If you have platies, you want red-tailed sharks; if you have plastic plants, you want living ones; if you have a five-and-a-half-gallon tank with one betta, a black neon, two catfish, and an algae eater named Glom, you want four 1,000-gallon tanks, covering every wall of the living room, which reproduce—exactly—the streams of Southeast Asia, Lake Tanganyika, the Amazon River, and the Great Barrier Reef. I began to see, after reading the advertisements and articles in *Tropical Fish Hobbyist* and monitoring my own rising lust for bigger tanks and weirder fish, that underneath the patina of interest in aquatic biology lies the true passion of the aquarist, a peculiar but intense form of greed. It was clear that I'd chosen the right hobby. And it occurred to me that there was no point in continuing to fool around with pet stores. I should follow my avocation to its ultimate manifestation. I went to Baltimore.

There is no greener grass, pisciculturally, than Baltimore. It's the home of the National Aquarium in Baltimore, which is so big that you don't walk around it, you walk around in it. Its two big tanks are rings, and you're in the hole in the middle of the rings, with fish swimming in circles around you. My aquarium holds five and a half gallons of water and five fish, and cost me, including light, filter, plants, and rocks, about $50. The National Aquarium has a million gallons of water and almost 5,000 fish, and it cost Baltimore about $20 million. If the guy who runs the National Aquarium gets tired of fish, they get rid of him, not the aquarium.

Naturally, I approached the aquarium not as a layman but as an aquarist. So what if its circular Open Ocean Exhibit has 220,000 gallons of water and sand tiger sharks the size of my Volkswagen? It's still an aquarium. For instance, in my tank the betta chases the other fish and will sometimes nip them. The same thing happens at the National Aquarium. One day a brown shark ate a sheepshead and started writhing in indigestion. Attracted by the motion, one of the sand tigers nipped the brown—in a manner of speaking. He eviscerated the smaller shark with one bite. You can argue that retrieving a four-foot disemboweled shark from a tank that still includes his disembowler is fundamentally different from picking out a battered inch-long black neon from a home tank with your fingertip. And that may be so. A four-foot fish cannot easily be buried in the yard, and it does seem wrong to say "Bye-bye, fish" to a dead shark.*

There are a few other differences. When you feed the sharks, you stand on an unprotected catwalk over their tank and waggle fish on the end of a pole in between their jaws. You don't want to fall in. In feeding my fish, I've experienced a mild ennui, but never outright fear. Then there's the cost of the food. The National Aquarium spends $100,000 a year on krill, lettuce, peas, herring, broccoli, oysters, clams, shrimp, and assorted other forms of fish chow. The aquarium also spends $100,000 a year on salt. The water—plain Baltimore tap water—goes for $25,000 a year.

And there's the filter. Aquarists, by default, must love technology as much as they love fish. An aquarium isn't like a hamster cage, which just keeps the hamster from

---

*In fact, the National Aquarium doesn't have a yard (that I could see), which may be why they send their dead fish to the University of Maryland pathology laboratory.

escaping to die between the walls and ruin dinner parties. An aquarium maintains creatures from another world; it's like a reverse submarine. And what keeps it going is the filter. For the home aquarium there are under-gravel filters, sponge filters, box filters, outside-the-tank filters, all the kinds of filters you could imagine. But the filtration system of the National Aquarium is to the home filter as the space shuttle is to a paper airplane. There are no plastic tubes in this set-up. I'm talking twelve-inch pipes, 45,000-gallon holding tanks, atomic absorption tests for water quality, mass spectrographs, refractometers, computers, lots of big, big pumps, and two rooms—one huge, with pipes and pumps and tanks painted in bright colors, like the Pompidou Center turned outside in, the other steamy and small, where the water flows over plastic cylinders on which grow bacteria that digest fish waste. I'm talking about a system that circulates all 555,000 gallons of water in the aquarium's two big tanks in only ninety minutes. To put this in terms we can all appreciate, if this system were hooked up to my tank, in an hour and a half it would filter the water, and probably the fish, 100,909 times, approximately.

Maybe this is something that only appeals to those of us who read articles like "Crustaceans in the Home Aquarium (Part 6—Mantis Shrimp and Barnacles)" and wait eagerly for Part 7. Perhaps there exists, somewhere, a person able to stand in the filter room of the National Aquarium and be unmoved. On me the effect was overpowering, and unexpected. I no longer covet a larger tank, for the same reason that mystics who've experienced nirvana don't apply for American Express cards. Once you've seen the Platonic Ur-Aquarium of Baltimore you're forever spoiled for the fifty-five-gallon fully equipped tank and stand now on sale for $144.99 at my pet store (a good deal, even if I am no longer interested in it).

This isn't to say I've abandoned my aquatic fantasies altogether. What happened during my spiritual experience in Baltimore was that I learned that some of the feeding of fishes in the 335,000-gallon Atlantic Coral Reef ring tank is done by scuba divers. I saw them, floating with the fish, petting them, handing out bits of food. I also learned that these divers are almost all volunteers. Inevitably, the following thought occurred to me: If I can persuade my wife and children to move to Baltimore; if I become a certified diver; if I take the special course for volunteers—then, although I may never be able to get a 335,000-gallon aquarium for my living room, I'll be able to get *in* one.

Mother Goose Biology

There's a reason people hit alligators over the head with sticks. It's not genetic. It's cultural. In fact, it's literary. But first the alligator story: About two years ago I was on a nature walk around a pond in the Everglades—looking at egrets, hawks, and roseate spoonbills in the company of people who said things like "The common egret has black feet; it's the *snowy* egret that has *yellow* feet, sweetheart." We happened to pass by a big alligator and the level of the conversation took a sudden drop. (The same people said things like "That's a *big* alligator.") The ranger leading the walk told us that the year before she had come on a young woman beating a similar-sized alligator over the head with a stick. The alligator hadn't eaten the woman's dog, or bitten off the hand or foot of a loved one. The woman liked the alligator. She was just trying to get it to open its mouth so her boyfriend could get a good picture.

Next Ernest and Celestine: They are, respectively, an adult male bear and a young female mouse who share an apartment in a city I take to be Paris. (There is an alligator connection here.) They appear in several children's books by Gabrielle Vincent, which I bought for my daughters. The Ernest and Celestine books are very sweet, but, biologically speaking, there's something horribly wrong with them. I don't mean the fact that a bear and a mouse live together in an apartment. What I'm talking about is the fact that they are a bear and a mouse. What is their relationship? One might think at first, since Ernest is very fatherly, that they are members of the same family, or at least of the same species. My sixteen-year-old nephew made this assumption when he started to read one Ernest and Celestine story to my older daughter. Then he became completely confused by the drawings. "I don't know," he said of Celestine. "It looks an awful lot like a rat to me."

His comment was what opened my eyes (I didn't figure all this out by myself). I realized, in a way that I hadn't before, that she was a rat—or a mouse, anyway. Why wasn't she a bear? Who was Ernest? Then I thought of the song about the fly marrying the bumblebee (the ultimate interfaith marriage), of all the stories in which squirrels, rabbits, cats, and foxes peacefully co-exist, and, by what seemed to me inexorable logic, of that woman smacking the alligator. Kids' books did it to her. That was my conclusion. She hadn't just assumed that all of Florida was part of Disney World— a common enough mistake. She was the victim of countless silly children's stories (some purveyed by Disney, to be sure) that had turned her sense of the natural world upside down. Maybe she didn't believe that the alligator was a postman who was married to Rosie Spoonbill, the village gossip. Maybe. But she didn't believe it would bite her either.

And no wonder, if she was brought up on the books my children read. Think of what Ernest and Celestine do to biology. If it were just the two of them, that would be O.K. All species have eccentrics. But their world is infested with adult bears and young mice in family units. When Ernest and Celestine visit a more established household (they themselves live in reduced circumstances), we don't suddenly see little baby bears running around, or Mama and

Papa Mouse. We see two adult bears, male and female, presumably married, and a brood of mice for Celestine to play with. Children? Are we supposed to believe that the mice are the bears' children? Or is some odd kind of symbiosis going on—interspecific foster care? If so, when the mice leave their children on the doorsteps of bears, how do they steal away the cubs? Perhaps (I haven't read every Ernest and Celestine book, so I can't be sure) Gabrielle Vincent has constructed a world like that of H. G. Wells's *The Time Machine,* in which ugly underground people, the Morlocks, create a comfy world for pretty, childlike, aboveground people, the Eloi, whom they eat. Could it be that the mice aren't the bears' children, but their snacks?

As any parent knows, the Ernest and Celestine stories aren't the only "alligator beaters" (as these books are called in the trade, or should be) on the market. Another remarkable series is produced by Richard Scarry, the James Michener of kidlit. Richard Scarry dispenses more information per page than any other children's author. I think his books look cluttered, like a house too full of knickknacks, but children, perhaps because they have so much room on their mental shelves, love them. In all his books, all animals pal around with all other animals. A family of cats has, as a friend, a worm, who, in a triumph of impossibility that even I find irresistible, wears a shirt, a pant, and a shoe. (He ties his own shoelace, which I leave for you to imagine.) The classic Richard Scarry scene has a fox, a rabbit, a dog, and a cat enjoying themselves, with no interspecific conflicts, on the same swing set. I suppose it's pointless to complain that his creatures (except for the worm) are so denatured that they have no identities and have become interchangeable cuddlies. But when he puts a lady pig in a butcher shop slicing baloney, and hanging next to her is a ham— that's sick.

In the sea of happy furriness that greets the parent shopping for kids' books, one volume is *hors de classe,* if not *sui generis* (it's hard to tell with these foreign languages) in its sins against taxonomic integrity. The book is *Stuart Little,* by E. B. White. There is no simple matter of cats taking earthworms for car rides. In *Stuart Little* a human family gives birth—or perhaps I should say gives rise, since this is more like speciation than reproduction—to a mouse. (Presumably this story was the inspiration for Ernest and Celestine.) The Littles (who are actually full size) take the mouse to their bosom, being careful not to crush him, and he becomes a valued, if small, member of the family. I can't say it's not a good story. I ended up rooting for the mouse, as I'm sure everyone does. But I can't help wishing that Mr. Little, when he was first presented with a rodent in swaddling clothes, had been as honest as my nephew, and had said to the nurse, or Mrs. Little, or whoever was around, "I don't know, it looks an awful lot like a rat to me."

It's not so much anthropomorphism that I'm opposed to. I like a good talking duck as well as the next daddy. I just want the duck to be a duck, not a Smurf in feathers. It's not an impossible trick. Beatrix Potter carried it off, with foxes and rabbits as well. In one of her stories, Jemima Puddle-duck is bamboozled by a smooth-talking fox into providing the fixings for her own roasting, but is saved by the farm's collie dog and two foxhounds. This is no biology text—the fox talks—well—and he even has a kitchen, or some place to roast his ducks before he gobbles them up. But look at the personalities of the animals. The fox is unscrupulous. The duck is an idiot. The collie is a responsible sort, and the foxhounds are enthusiastic, but not too bright. Enhancing the realism, the foxhounds, once they save the duck, eat up all her eggs.

Mother Goose also contains some sound animal stories. Three blind mice getting their tails cut off with a carving knife may be bloody and a bit surreal, but it's nowhere near as bizarre as *Stuart Little*. And when Little Miss Muffet is scared off her tuffet, I think no student of either children or spiders could object. (We'll let the whole matter of what, exactly, a tuffet is rest for now.) Furthermore, the stories that do justice to animals also do better by the human beings that I assume the animals are supposed to represent.* Aesop, though not usually thought of as a children's writer, could well be used to give young people a solid grounding in both animal and human behavior. In one of my favorite fables, a wolf tries to convince a lamb that it has done something wrong, and should therefore be eaten. The lamb correctly and skillfully argues its innocence. Then the wolf decides to stop wasting its time and says, "Well, you may be a pretty slick talker, but I'm going to eat you anyway." I've had similar exchanges with editors.

Finally, there are the tales collected by the Brothers Grimm, which are satisfyingly full of the duplicity, hunger, and violence that characterize life in the woods, the playground, and the office. The best may be Red Riding Hood, if you read the Grimms' tale and not one of the watered-down versions in which the wolf merely scares the grandmother and the girl. According to the Brothers Grimm, the wolf eats both of them. They're liberated when a huntsman slits the beast open, then fills him with rocks and sews him back up. And the grandmother isn't brought trifling sweets, but a cake and a bottle of wine, a gift that would be more to the taste of the grandmothers I know. A lot has been written about this story, and I'm not going to start a new analysis of the psychosexual currents running beneath its

---

*This brings up the question of why so few books for children are about actual, unfurred people (*Homo sapiens sapiens*). I suspect the reason is that even the smallest child knows from personal experience that there are limits to the sweetness of even the nicest adult of its own species. On the other hand, kids will believe anything you tell them about bears and mice.

surface. I like it because it contains a wonderful poetic statement of the workings of evolution. We all know that natural selection forged the wolf's teeth into tools to rend and tear flesh. But that's a ponderous way to say it. How much more felicitous to have the wolf himself declaim, in response to Red Riding Hood's innocent wonder (the ultimate source of all scientific inquiry), "The better to eat you with, my dear." Darwin himself must have loved that line.

The Urge for Going

There is a famous goose song called "The Urge for Going." Joni Mitchell wrote it and Tom Rush sang it (he probably still does) huskily, sometimes not quite on key. In it, Canada geese appear as romantic figures, as they do in most goose music. They fly in V-formation in the crisp autumn air, and their flight, squarely in the tradition of the elegiac goose lyric, is a symbol of change, of loss, of wanderlust. We all know these geese—going south, going north, always going somewhere. These are what we call songwriter's geese.

They are not, however, representative of all Canada geese. Some of the others waddle along on the ground with an air of utter pragmatism. And honk. Unmelodiously. Loudly. When Felix, in "The Odd Couple," clears his sinuses, the noise he makes isn't precisely the same as that made by a waddling Canada goose, but it has precisely the same amount of charm. These other geese will fly if they have to, but only to avoid freezing or starving to death. If they get the chance to trade their long migratory past for a park pond and white bread year-round, they snap it right up with their greedy little bills.

Some of these geese have moved to Greenwich, Connecticut, a town that is near the top of the list on all three of the important criteria for residential real estate—location, location, and location. It has vast lawns, parks, ponds, a high demographic profile, and Long Island Sound. It never gets too hot or too cold, and the roughly 5,000 geese that have settled there apparently blithely ignore the whispers around town that they aren't Republicans. Of course, the truth is that there are Democrats in Greenwich, and not all of them are geese. And although many people in Greenwich don't like the geese, this has nothing to do with politics or social standing. It's the kind of reason that never gets into goose songs. Droppings. Five thousand geese can drop a lot of droppings on the beach, the estate, and the fourteenth green. And when they do—that, if you'll pardon the expression, is when the scat hits the fan.

Greenwich recently hit the newspapers and the evening news because of its goose droppings. It had gotten to the point where the park department had to borrow the yacht club's ceremonial cannon to shoot blanks to scare the birds off the beaches. Even the CBC (Canadian Broadcasting Corporation) called town officials for a radio interview. What they wanted to know was why these "Canada" geese weren't coming home.* The flocking instinct of journalists is well known, and I saw no reason not to follow in everyone else's path (being careful where I stepped, of course). I went first to Byram Park, one of the places the geese frequent. It's a nice park—almost as nice as the property around some of the town's middle-range houses. The geese were there, on the grass of the ball field. They were also in Bruce Park, near the tennis courts, repetitively bobbing their heads up and down as they plucked the grass (geese, like buffalo, are grazers). They were pretty, with their black necks and that splash of white on their cheeks (I'm not sure that geese

*A sign, I take it, that at least some Canadians are starting to take themselves less seriously.

have cheeks in the strict scientific sense of the word, but I don't know what else you'd call them). And they were pleasant to watch. But—I don't want to be coy about this, so I'll say it straight out—there was goose doo all over the place.

If it were only Greenwich, only Greenwich would care. But these stay-at-home geese aren't just a few renegades. There are nonmigrating geese all over the country. They're moving into the parks and onto the golf courses and lawns. They're precipitating an environmental crisis of unique proportions, a competition for habitat (and nice habitat at that) that could be the severest test yet of America's love of wild creatures. Among the places afflicted with nonmigrating geese are Massachusetts, Connecticut, New York, New Jersey, Pennsylvania, Virginia, Minnesota, and the cities of Seattle, Denver, Cleveland, Nashville, and Toronto. (Some of the geese have stayed in Canada.) Minnesota has an urban goose population of about 30,000 to 40,000. Connecticut and New York have about 10,000 between them.

What has happened is that instead of going back and forth from Canada to the South (back and forth, back and forth—I'm sure the geese hated it), some pioneering birds found climates where it never got too hot or too cold, and where there were lawns to eat, and, even better, people to feed them. Now, according to Kathryn Converse of the U.S. Fish and Wildlife Service, these nonmigrating geese hop from park to estate, looking for new nesting sites as their population grows. This is how one gets geese: Two geese stop in the yard in the spring, checking out your pond as a place to raise children. Two geese in the yard look cute. As everyone knows, the natural food of geese is stale bread, so you give them some. In Greenwich, maybe you have the cook buy an extra baguette for the birds. They eat. They stay. They reproduce. More geese. More cuteness. More goose poop. In three years there are fifty geese—permanent geese that don't even know where Canada is—and everyone in the neighborhood is wearing rubber boots.

Converse, who provided me with most of this information—I am not, personally, a goose expert—knows quite a bit about geese. She did her Ph.D. thesis on nonmigrating geese in Fairfield County (Connecticut), where Greenwich is, and neighboring Westchester County (New York). She saw the human misery firsthand—houses where people had to have two sets of shoes (two complete sets of shoes!) because of slippery grass. Ornamental ponds where the water had turned green—and I don't mean the emerald green of the Caribbean, I mean goose-doo green. She followed geese around—1,000 individually marked geese—for three years. And she got to know them and their desires. Geese like water to land in and sleep on because while they're on it they can see predators coming. They like grass to eat, particularly short grass, because the new shoots have the most protein. They like to nest on islands or peninsulas. They like water that doesn't freeze in the winter, or doesn't stay frozen for long periods, like the bays of Long Island Sound. And they love people who feed them. If you were to design the ideal wildlife management area for geese—it would be Greenwich. Geese, in other words, like exactly the same environment people do, and since they don't have to buy houses or pay taxes, they can live wherever they want.

You might think that when the geese get to be too much, you can just run them off the property. And it is possible to get rid of geese, but it's not easy. First, never feed them. Then, make sure nobody else within a mile of you feeds them. Build barriers around the ornamental ponds (forget what they look like). Finally, harass the geese, incessantly. Persistence is the key. Nastiness and meanness of heart are also useful. Converse says that in the course of her research she would find one golf course with geese, and another with none. The goose-free course would have achieved its idyllic state by having employees on golf carts charge into flocks, shouting unfriendly things at the birds, every day. Geese, when they know they aren't wanted, won't stay. The difficulty is in convincing them that you don't want them.

It is this obtuseness, this complete failure by geese to take a hint, that's bound to drive goose-ridden communities to violence. The current balance, with the geese happy and the people unhappy, can't last. The birds have been smart enough so far to settle in residential areas where there's no goose season. But the nation that taught the passenger pigeon and the buffalo their lessons isn't going to let the fish and game laws stand in its way. I see a lot of 12-gauges coming down from a lot of attics and snuggling against a lot of homeowners' shoulders. I see a lot of intentional disregard of the Migratory Bird Act of 1918. I see a lot of Labrador retrievers that have spent their lives chasing tennis balls and getting squished between suitcases in the back of station wagons finally getting the chance to retrieve real, dead geese—by the thousands.

Nor will the geese be the only victims. And this is the real tragedy—the birds are going to spoil it for all the other animals. I'll explain why. First, I think we can assume, *a priori,* that there are huge numbers of people whose love for other living creatures is founded on song lyrics. Seeing or hearing about other creatures at a distance, given the right chord progression, can make one inordinately fond of them. Because the average lover of wild things doesn't know the objects of his affection intimately, it's easy to spring to their defense. Take coyotes—they eat the odd pet here and there, but they haven't made a habit of it. Maybe they do eat lambs, but then you know those shepherds, always crying "Coyote!" As for grizzlies, there's no doubt that they kill people, but I've detected, in conversations with the ecologically sensitive, a suspicion that they don't kill *good* people. They kill people who sleep in the wrong places, or carry salamis into the woods, or use deodorants and perfumes in Glacier National Park. Like test pilots, lovers of nature feel that there's a kind of hikers' right stuff that will see you through your confrontations with the natural world. (Most of the potentially anti-grizzly hikers, are, of course, already dead.)

As for the other wildlife one is likely to encounter between Greenwich and Wall Street, pigeons tend to stick to public buildings and statues, and although raccoons will steal your garbage, they don't mess around with your lawn. Rabbits are sensitive to insults. And rats—well, rats go after renters. But geese, geese are birds of a different feather. And once you get to know a goose, up close and personal, it's difficult ever again to get misty-eyed about the wonders of nature. An animal that makes that incredibly nasal noise, that's primarily interested in its own comfort, and that would rather live on handouts than migrate for a living is clearly no better than a human being. If that's true of one wild creature, it's probably true of the whole lot. You see the inevitable conclusion: If they (all other living things) are no better than we are, and if it comes down to a fight over the same lawn—*Blam!*—or pond—*Blam!*—or national park—*Blam! Blam!*—why shouldn't we be the ones to get it?

The Sign of the RAM

I have just had my horoscope done by a computer program. This wasn't an experiment in artificial intelligence. I wasn't asked to read a sheet of paper that informed me that I was "a free spirit" and to determine whether this bit of wisdom had come from a person or a computer. That would have been an impossible test, since all horoscopes read as if they were written by computers. I simply bought a computer program called Deluxe Astro-Scope, plugged in my date, time, and location of birth (latitude and longitude were required—no Mickey Mouse here), and waited while the disc drives whirred, Alan Turing* turned over in his grave, and the printer spewed forth ten pages of planets, aspects, midheavens, and modalities. It even told me, in a roundabout way, my sign—Taurus.†

---

*Turing, whose sign was Cancer, was an early theorist of artificial intelligence.

†What the computer didn't tell me, but I knew anyway, was that there is a homonym for my sign, "torus," the term in solid geometry for the doughnut shape. This fact has given me a perverse yearning to dress up as a glazed doughnut, walk into a California singles bar frequented by mathematicians, and say to some blonde who looks as if she knows her solid geometry, "Hi, I'm a torus."

The printout contained an enormous amount of exotic information that's no doubt of great value to the professional astrologer. I can now tell you that my Neptune is in House Twelve, my moon in conjunction with Mars, my Jupiter in Aquarius—and my bill is $295—for the floppy disc and accompanying documentation. Insight isn't cheap. The program provided me with some startling pronouncements on my character. It said, for instance, "You have an unyielding nature, which makes it very difficult for those with alternate opinions to co-exist with you." This, of course, is absolute nonsense.

Mine isn't the only astrology program; it's one of many. Depending on your point of view, computers have invaded astrology, or astrology has invaded computers. There are at least two companies devoted to astrological programs: AGS Software of Orleans, Massachusetts, from which I purchased Deluxe Astro-Scope, and Matrix Software of Big Rapids, Michigan. The emphasis in their catalogues is on professional astrology. In the Matrix brochure a headline trumpets "Make money with your home computer!" And AGS notes that "you can make Electronic Astrologer Astro-Reports consistent moneymakers in a horoscope calculation service." That's not why I ordered the program, but I was glad to hear it. If the writing business goes sour, I'll be more than happy to tell you (to borrow the typography of the printout) whether your horoscope is dominated by FIRE, EARTH, AIR, or WATER—for a small FEE.

The programs Matrix and AGS offer can do almost anything. With them you can not only cast an individual horoscope but also check the compatibility of two people, do astrological research, biorhythms, numerology, and tarot readings. You can even do a little sexual astrology for consenting adults. The brochure from AGS touts their Deluxe Sex-O-Scope as providing "a playful, witty, R-rated description of romantic and lovemaking styles and preferences.

Does not include explanatory pages—these we leave to your ingenuity!"

My mother warned me about people who combined sexual innuendo and exclamation points, so it was with some trepidation that I called AGS about the program I wanted. I needn't have worried. My conversation with the woman who answered was thoroughly official. Instead of asking, alluringly, "What's your sign?" she said, "What's your operating system?" I wish I could say that when I answered "CPM" she said, "Ah, you're business oriented, a pioneer, with a large library of free software," but she didn't. She asked me about my RAM, not Aries, the sign of Bismarck and J. P. Morgan, but Random Access Memory—that RAM, the sign of Stephan Wozniak and Steven P. Jobs.

As to the program she sent me, I can't fault it. It provided better, or at least more, advice than I ever got from the newspaper. At times it appeared incredibly perceptive, as when it pointed out that I was "courageous and daring." Then it would go ludicrously off the mark, describing me as "vain and lazy." Still, it was nice to have the feeling that someone was taking an interest in me, talking to me, about me, even if he was saying nasty things. The astrological second person ("You are tall, dark, and handsome," "You will become amazingly wealthy, today") is irresistible in its illusion of intimacy, whether it comes from a computer, a newspaper column, or an astrologer in the flesh. Even when you know perfectly well that a million other people are reading "You are a deeply passionate person who needs endless love—from lots of different people," it can still feel as if someone has finally understood you.

The only problem with astrology is that it's all hooey. I didn't make this up just to be mean. I got it out of the *Encyclopædia Britannica,* to which I often turn for guidance when I'm not reading my horoscope. The encyclopædia said, in a tone I thought was a bit harsh, that after Newton astrology became "scientifically untenable" and, in the West, "more and more fraudulent." It called it a pseudoscience and said, "Modern Western astrology, though of great interest sociologically and popularly, generally is regarded as devoid of intellectual value." In other words—hooey. What, then, does it mean for the culture as a whole that computers are being put to uses "devoid of intellectual value"? I think it's very good news.

The advent of computerized astrology marks the intrusion of silliness into the halls of science. The computer is the closest most of us get to a scientific advance. It's an electronic icon, the reigning trinket of twentieth-century technology. It's one thing to use such a device to pretend you're commanding the Starship *Enterprise* and trying to crush the Klingon Empire (at least that's *science* fiction). But it's another to force the computer to play handmaiden to the occult, to irrationality and superstition. That's a scandal, an outrage. As Henry Higgins would say: How simply frightful! How humiliating! How delightful!

It's not that I rejoice in seeing the poor computer dragged in the mud of the zodiac. It's just that I've been worried by dire predictions that computers will dehumanize us, that they'll take over our lives, suck the juice out of them, and leave us nothing but bits and bytes. Computer astrology seems to put the lie to these claims. Astrology may be dumb, but it's human. If even science writers are sitting around forcing computers to do silly, irrational, and useless things like cast the astrological charts of Prince Charles and Howard Cosell,* then we are as likely to end up dominated by the chill, restrictive logic of the computer as a Taurus is to change his mind.

---

*Prince Charles, a Scorpio, has "a preference for rich, elegant surroundings & possessions" (sic). Cosell, an Aries, is "quite tolerant of others' faults," as everyone knows.

One has to remember, when considering the potential dehumanizing effect of computers, that being human isn't always such a noble thing. Part of being human may be caring for families and friends, reading (or writing) great literature, and going to India to help Mother Teresa. It is, however, equally, if not more, human to bet on horses, philander, read your horoscope, play games, and try to make money by selling astrology programs. All these activities, and others, remain possible, if not easier, with computers. There are programs to handicap horses, keep bowling scores and averages, do biothythms, throw (or perhaps I should say compute) the I Ching, and teach you how to win at blackjack. There's Deluxe Sex-O-Scope. On the same machine you can go shopping, trace your family tree to Prince Charles (or Howard Cosell), and send electronic love notes to compatible computer owners. And there are the games, from Space Invaders to Bible Baseball. In one catalogue of software for Apple computers there are two pages of general science and more than forty pages of games. Obviously, computers aren't turning people into humdrum machines. People are turning computers—as they have every other bit of technology, from the internal combustion engine to the machines (I know they exist, even though I've never seen one) that make rubber dinosaurs—to their own frivolous and irrational pursuits.

It's true that I'm only talking about home computers, so I don't want to wax overly optimistic. Until we get the computers at the Pentagon and the Kremlin I-Chinging and doing horoscopes for every person that ever lived, we're not really safe. But there are teenage hackers who can set that up for us. And if you can't abide astrology, think of it this way: a computer that's busy worrying about the difference between Gemini and Virgo isn't fooling around with your bank account.

Remember, these words come to you from "a perfection-ist" with "a deep and inquiring mind" who, although he "loves practical jokes" and has "a tendency to be shallow," is nonetheless graced with "cool logic." I couldn't have put it better myself.

There's This Tribe . . .

My wife, when she was studying psychological theory, would often come home and regale me with stories of how a child's upbringing could damage his psyche—breast-feeding to age twelve causing uncontrollable eye twitching, or toilet training with rewards of M&Ms leading to idiopathic colitis each Halloween. Being naturally argumentative, I always tried to come up with a counterexample. Well, I would say, the !Kung Bushmen of the Kalahari breastfeed until the midlife crisis—the child's—and there's nothing wrong with them. And what, I would ask, did she have to say about the Yanomamö of South America? I happened to know that they didn't celebrate Halloween, nor did they have M&Ms—let alone toilets.

It could be that our arguments on matters of human nature were the result of my own severely flawed personal character. But I don't think so. I blame the fights on anthropology. In no branch of knowledge is it so true that having a little of it can be a dangerous thing. And I know of no other field where so many know so little about so much. Some of us use our knowledge responsibly. I tend to use my own vast store of information on the Gururumba to speak for the dignity and diversity of all that is human. Other people, from what I can see, use the little crumbs of information they've garnered from Anthro 101 to make trouble in what would otherwise be pleasant conversations.

We all have our favorite tribes, peoples who can be thrown up to devastate a conversational opponent who thinks he has just explained why men (or women) are genetically designed to take out the garbage. "Ah," you say, "but among the Gazonga there is no garbage!" The classic choices are the !Kung, the Ik, and the people who stretch their lips out with little round discs (which I've always thought were distressingly reminiscent of clay pigeons). This trio is known, at least to me, as the good, the bad, and the ugly.* But there are many other possibilities, like the Hopi, the Pygmies of the Ituri forest, the Kwakiutl, the Inuit, the Yir Yoront, not to mention cross-cultural studies.† If you missed anthropology in college, you needn't fear that you'll lack examples. Public television recently ran a Tribe of the Week series (it was actually called "Disappearing World"), allowing their upwardly mobile audience to get acquainted with peoples who were headed in the other direction. (I think it's fair to say that disappearing from the face of the earth is the *ne plus ultra* of downward mobility.) I'm convinced that somewhere in the world there's a tribe for every point of view.

If you'd like to argue in favor of adultery, for instance, you're in good shape. Sex has always been a mainstay of anthropology. We do have it in our own culture, but it always seems better in places like Micronesia. Variety is the spice not only of life but of ethnography. I suggest you read

*The !Kung are the hunter-gatherers who treat their children well and make no mortgage payments. The Ik are the people who degenerated into pure selfishness when faced with starvation. The clay pigeon people are familiar to anyone who has ever looked through any copy of *National Geographic*.
†Who could resist the comparison of the Machiguenga, of the Amazon rain forest, with the Parisians, of France?

Ulithi: A Micronesian Design for Living by William A. Lessa, published in 1966, which I found in a carton of my wife's old books from college. There is an unclothed girl on the cover, and if you zip directly to chapter seven, "Sexual Behavior" (I had to do it, I was researching this column), you'll find an account of the holiday of *pi supuhui*. This Ulithian phrase is loosely translated as "a hundred pettings." The idea behind the holiday is that each person pairs off with somebody of the opposite sex and heads for the bushes. Spouses can't pick each other. The best part about *pi supuhui* to my mind is that it's not set for any specific day. It's held whenever anyone suggests it. I envision the Ulithians calling for *pi supuhui* in much the same way that undergraduates shout "Food fight!"

Sometimes it's hard to figure out what point of view a tribe is meant to support. The Jalé of New Guinea are cannibals (or were, as late as the 1960s) who clamped the eyelids and lips of their victim (or meal) shut with bat wing bones so the soul wouldn't get out and give them indigestion. Although the Jalé had some religious reasons for the way they treated their enemies up to the point where they were well done, the reason they ate them was not sacred but profane. People taste as good as or better than pork. I guess the Jalé could come in useful sometime in an argument about who has—or is—the best barbecue.

There are two ways to look at argumentative anthropology. You can use it, or you can deplore it. I favor using it, which means you should try to be the first one in a conversation to bring up a primitive people. This gives you a definite edge. If your opponent (I confess to viewing all conversations as contests of one sort or another) is first, he has the advantage. You are put in the position of a poker player responding to a bet: "I'll see you your circumcision rite and raise you an exogamous marriage."

Of course you may be one of those people who value intellectual rigor, honesty, friendship, and a fair fight above conversational victory. I think this is a mistake. I know, however, from my anthropological background that it takes all kinds (the first principle of comparative ethnography). If you feel besieged by unscrupulous tribe mentioners, there are ways to defend yourself.

First, pay close attention to the way the tribe is described. If, in the course of a discussion on, say, the importance of mesquite for that smoky taste, or the role of raspberry vinegar in a good sauce, a professional anthropologist (or chef) were to mention the Jalé by name, and discuss their recipes in detail, you would know, at least, that you had a real primitive people to deal with. This is a rare occurrence. Most of the people who talk so much about these tribes aren't anthropologists at all, and they often don't know what tribe they're talking about. They begin with a long, erudite sigh and a sly look, and then they drop it on you. "There's this tribe . . ."

Don't you believe it. Ask who this tribe is, and if the name isn't forthcoming, express grave doubt as to its existence. I know this seems harsh, but people are always mentioning tribes they think their professors mentioned fifteen years ago, but which, in fact, are no more real than the Gazonga, which I must admit I made up. The Gururumba, on the other hand, are real, which proves that truth is more sonorous than fiction. I'm not the only one to make tribes up. Ludwig Wittgenstein, the philosopher, also did so, but he was honest. He would write, for example, "Imagine that the people of a tribe were brought up from early youth to give no expression of feeling *of any kind.*" (Then, presumably, they would *never* use italics.)

Something else to look for in tribe aficionados is a slightly tarnished version of the myth of the noble savage. The tribes are, for many of us, the human equivalent of herbal shampoo. People who live in the woods and don't have designer clothes, or even clothes, are assumed to be natural, and therefore better. This isn't necessarily true. Consider the mode of dress of the Jalé. True enough, penis sheaths are morally superior to Jordache jeans and pinky rings. But Jalé men also wear piles of hoops around their waists, which is ridiculous. Furthermore, for every good tribe, there's a bad one. If someone is trying to tell you how much better, more wholesome, and *real* tribal life is, as opposed to the cocaine-ridden treadmills of the better neighborhoods in the Free World, remind him of the Yanomamö. Not only are they devoted to war, but their favorite nonviolent recreation is to spend the afternoon sitting around the rain forest blowing green hallucinogenic dust up their noses.

You needn't counter with a tribe of your own, if you're at a momentary loss for anthropological words. And even if the tribe in question is real, you aren't yet lost. How do we know that the tribe was telling the truth when they told the anthropologist that Coyote stole the secret recipe for boiled maize so the people wouldn't starve when the pine nuts failed? I wouldn't bring up the possibility of mendacity among primitive peoples if it weren't for the recent Margaret Mead brouhaha. You may recall that Margaret Mead came back from Samoa with the news that teenage girls there got to sleep around before they were married and their daddies didn't take their T-birds away. Then, a few years ago, Derek Freeman said this wasn't true, that the parents would never have let them go out riding with boys in the first place, even if they had cars. He said that Mead's informants probably told her fibs to tease her, a form of behavior called by the Samoans *tau fa'ase'e,* or "giving her the business" (my translation). I don't know who's right in this argument. But it does give one pause. If these tribes lied to Margaret Mead, they'd lie to anybody. It may even be that making up stuff to tell anthropologists is the main entertainment of primitive peoples:

First Jalé man: Let's tell him we eat people.

Second Jalé man: Great. And I'll get a bunch of those hoops . . .

If all else fails and one is faced with a real, unarguably extant and truthful tribe, there is a final counterargument, which relies on brutal logic. I'm telling this out of pure altruism, since my wife reads these columns and I'll never again be able to use a tribe to argue her out of a position on what's psychologically healthy. Let's say the argument from the tribe side is that this tribe breast-feeds forever and they're O.K. The counterargument is "Who says they're O.K.?" As professional anthropologists know, just because somebody, somewhere, does something doesn't mean it's good. This point was made most succinctly by a friend's psychoanalyst. My friend had just finished talking about some tribe, probably trying to say that if they didn't need to pay $80 an hour to get along, maybe he didn't either. But were they getting along? "What do we know about these tribes?" the psychiatrist said. "They've never been analyzed." Then, I suppose, he handed over that month's bill.

The Man with No Endorphins

A neurochemical vignette: There's a man, running in the rain, wearing loafers, in Baltimore—it's me. I'm not happy. I'm late for a talk on opiate receptors in the brain because I've been in traffic, then in the new Baltimore subway (which I must say works a lot better than my central nervous system), and finally in the rain, risking life and limb and ruining my shoes. I desperately want to learn more about endorphins and enkephalins, the brain's own opiates, which are supposed to ease pain and produce pleasure, but I know I'm going to be late, and probably wet. (I can never remember: Do you stay drier going faster, or standing very straight under the umbrella and taking tiny steps?) My heart is pounding, my anxiety rising, and my toes are damp. One thought is foremost in my mind. I'm thinking: "Where are my endorphins when I need them?"

I've always been fond of neurochemistry; a field that brings you the opiates of the brain is hard not to like. But neurochemistry hasn't treated me well. I'm sure some brains manufacture these great chemicals, and that the people who have these brains experience runner's high and other pleasant effects. But as near as I can tell my brain doesn't *do* endorphins. When I run I get shin splints and twisted ankles. The most I've ever gotten out of running, in emotional terms, was a momentary absence of anxiety, which I attributed to complete physical exhaustion. It was O.K., but it wasn't that different from being depressed. Runners tell me I never ran far enough. But I happen to know that earthworms have endorphins. Earthworms don't jog. And if invertebrates don't have to run to be happy, I don't see why a higher (or at least taller) vertebrate like myself should have to.

The truth is I'm not even interested in getting high. I'm not greedy. I was happy enough with the mild depression that followed running in the park to continue jogging for years. I stopped only because my twisted ankles refused to heal. What I'm really looking for is absence of pain, a certain kind of pain, which I'll try to define. There is traditional physical pain: cuts and bruises, having your head chopped off. I'm not talking about that. There is traditional psychological pain: Oedipus and Electra complexes, schizophrenia, anxiety neuroses. I'm not talking about that. I'm talking about another class of pain, which occurs in huge quantities every day in my neighborhood. This kind of pain is caused by computers, customer service personnel at banks, medical insurance and expense account forms, and airline baggage personnel, not to mention airlines. Let's call it First World pain.

I know that good people, when their taxes are due or their computers fail, realize that there are people in the world who have schistosomiasis, so that it would be incredibly selfish and insensitive to whine about capital gains or the loss of a great sentence when they still get to eat an unconscionably large amount of protein at dinner, which they don't have to share with blood flukes. Unfortunately, a lot of us aren't good people. A lot of us are bad. A lot of us are so wrapped up in our own little First World lives that taxes and computer failure seem, to us, to cause intense pain, to us.

Out of this sort of selfishness—I myself pay taxes, and have recently experienced computer failure which induced in me not only extreme pain, but guilt for not curing, or having, schistosomiasis—I went to Baltimore to the talk on brain chemicals. Other reporters, I'm sure, were at the conference to report to the public news that would affect their health and welfare, to educate them. I went because I thought I might learn how to find and use my own endorphins. No such luck. Most of the news was about pain messengers—not the people who bring bills and rejection slips in the mail, but peptides that relay the news of tissue damage to the nerve endings so they can send the news to the brain and the brain can cause the mouth of the person with the tissue damage (i.e., a finger that has had the door of a Coupe de Ville slammed shut on it) to howl in pain and indignation. The first messenger in this system is called bradykinin. Bradykinin, according to Solomon Snyder of Johns Hopkins, whose talk I was late for, is the strongest pain-causing substance there is. Fortunately, bradykinin antagonists have been developed to bind the bradykinin before it gets to the nerve endings. Since bradykinin apparently carries messages about arthritis as well as bruised fingers, the antagonist could be rubbed on an arthritic knee (this is all speculative) like old-time liniment, and it would scarf up all the bradykinin and stop the pain. What this could mean to millions of pain sufferers is obvious. What it means to me is that, in neurochemistry, sometimes it does make sense to kill the messenger.

However, to get back to my own pain, which, sad to say, is the subject of greatest interest to me, I have figured out why the endorphins don't work on it. The reason is that human beings were not designed, by evolution, to fill out tax forms, use computers, or fly on commercial airplanes. Now, we weren't designed to play the violin either, as I keep telling a friend of mine when he hauls out his fiddle and starts talking about Paganini.* But when it comes to fiddling we do have what biologists call a pre-adaptation: fingers. They didn't evolve for fiddling; they evolved, as we all know, to play the guitar. But if you've got good ones, they can be used to do hot licks on a Guarneri as well as a Gibson. There is, however, no similar pre-adaptation for dealing with the IRS or airline baggage personnel.

Skeptics among you may be mumbling that there's this tribe that's known for its incredible patience in hunting the dik-dik, but the dik-dik is a more appealing quarry than an old Samsonite suitcase, and the rain forest is preferable to the baggage carousel, even during fever season. In airports, not only do proselytizers try to convert you to obscure religions (opiates, opiates everywhere) but there are crowds of other people whose endorphins are also failing them, and who, for all you know, are about to pull Uzi submachine guns from under their coats and relieve their own pain by causing you to have some. Faced with this situation, the brain is at a loss. It doesn't recognize the pain you're undergoing as something endorphins can take care of, so it lets them sleep and leaves you to fend for yourself, unopiated.

It can be done, assuming, of course, that sooner or later the luggage stumbles through those flaps and the "There's-my-suitcase" neuron lights up. Sometimes that doesn't happen. And now we come to the airline personnel. At an earlier point in my life, when I was in the process of trying to summon up my endorphins, or at least distract myself, by crude means like exercise, I also tried breathing. To be precise, I tried Lamaze, not while giving birth, but in stressful situations. It didn't work, but it did provide me with a

---

*Paganini, on the other hand—on both hands in fact—had long spidery fingers, perhaps because of a genetic aberration called Marfan's syndrome. He *was* designed to play the violin.

tale whose moral is this: Ask not what your endorphins can do for you, but take your fly rod on the airplane with you.

One day a few months before the birth of our second child I was having a fit about some frustrating aspect of home improvement. My wife suggested to me that if Lamaze could get a woman through childbirth, maybe it could get a man through a conversation with an electrician. I tried it. I was never able to find the electrician, and the work is still not done, but I did use controlled breathing to call an airline and get information on flights to Great Falls, Montana. I stayed on the telephone, and kept my voice down for the whole twenty minutes, without benefit of medication.

I then went to Great Falls, Montana, with my wife, niece, first child, incipient second child, and, last but (I'm ashamed to say) not least, my prized fly rod. We were going to hike, fish, and look at dinosaur bones. You know what happened. I checked the fly rod and the airline lost it. I stood at the baggage counter, breathing—in, out, in, out, managing the pain—and filled out a form. (Have I emphasized forms as a source of pain?) We went to the hotel, and four days later I was told, in a telephone conversation (my wife was next to me with a cup of cracked ice, coaching me on my breathing), that the rod had been retrieved from Angola and sent to Kalispell, Montana, on the other side of the continental divide from the Many Glacier Hotel, to which I had been sent. The man from the airline, in the single most infuriating conversation I've ever had, replied to my calm, reasoned statement that I wasn't in Kalispell by telling me that that's where I should have been. There were more frequent flights to that airport, and, he said to me, as I stood there rodless and dumbfounded, the fishing was better over there.

I wasn't arrested for what I said into the telephone, but I could have been. Frankly, I don't see why, if these endorphins are going to be so fickle, we can't have more doctors around. When you're all worked up, there's nothing like general anaesthesia. I did eventually get the fly rod back, just as we were about to leave to go dig dinosaurs, and I used it a few more times that summer, back East. Optimists will see in this resolution a benevolent, smiling universe. Realists will see that the paragraph isn't over yet. On my last day of fishing (I didn't know it was my last day of fishing), I put the fly rod on the top of the car and, due to some faulty synapses, drove off. I heard a rattle, and in the side mirror I saw my rod leap into the air, do a barrel roll, and dive under the wheels of a pickup truck. There was nothing left but splinters. My endorphins, as usual, were nowhere to be seen.

A Serf in the Kingdom of Vegetables

And God said . . . let them have dominion over the fish of the sea, and over the fowl of the air, and over the cattle, and over all the earth, and over every creeping thing that creepeth upon the earth.

—Genesis 1:26

God didn't mention Japanese knotweed. Oh, there's that bit of hand waving about "all the earth" and some later references to herbs and fruits. But it seems to me that He left the question of where we stand in relation to plants somewhat vague. God was no dummy.

I used to like plants. But that was when I lived in an apartment. To me, a plant was a little green thing in an orange pot that needed to be watered, misted, and protected from mites. Plants were light, pleasant, undemanding, like salad, except they were still dirty. Then I moved to a house, a house that came with an overgrown plot of land that I saw, in a burst of romantic vision, as a yard. I discovered that plants in the wild are to house plants as the rats on the banks of the Ganges are to gerbils. The things that grew in my yard didn't need my protection; they didn't need anybody's protection. This is what a writer in *Horticulture*

magazine wrote about the things that grew in my yard—
"Viciously aggressive, rampant, and perniciously invasive,
these plants are best avoided."

I now believe that most outdoor plants are pernicious,
but the writer was describing two particular species, one
of which covered a good quarter-acre in my yard (the rest
was sumac, brambles, wild cherry, and poison ivy). That
plant is known, in the vernacular, as bamboo, Japanese
knotweed, Mexican bamboo, or "He That Eateth Up the
Yards of the Unjust."* It isn't real bamboo, however, but
either a member of the smartweed family, an apt name, or
closely related. Scientifically it's called *Polygonum cuspidatum* or *Reynoutria japonica*. I believe that plants are like
people in that if they have a lot of aliases, you should watch
out for them.

I once called the plant hotline of the New York Botanical
Garden in the Bronx—which is like poison control for lawns.
When I told them what plant I was fighting, the person on
the other end laughed. The same thing happens if you tell
a poison control volunteer you've just swallowed cyanide.
And for the same reason—it's an attempt to stave off black
despair with humor. I'm sure it seems silly that someone
could go on like this about a plant. But it isn't. I quote from
Weed Control in the Home Garden, a pamphlet published
by the Brooklyn Botanic Garden: "There are no sure-fire
cultural practices that will control Japanese knotweed.
Covering an infested area with two inches of asphalt in a
driveway is futile, as the new shoots push right through
the asphalt. Several layers of black polyethylene film tightly
applied to a leveled soil surface and covered with asphalt,
patio blocks, or stones may be an answer (but at considerable price for the average homeowner)."

You realize that the writer of that paragraph said that

*Maybe the just too. I can't speak for them.

if I spread polyethylene over a quarter of an acre and then pave it, that *may* be an answer.

In botanical terms Japanese knotweed is an "escaped ornamental." It's native to Japan, and according to one account was taken to England by a Belgian in 1864. It came to the U.S. in the late nineteenth century along with a relative from the island of Sakhalin known as giant knotweed, which is much the same, only bigger. Somehow, sometime, the thing got out of a garden or escaped from a hillside it had been meant to hold from erosion and made its way, over hill and dale, to my yard, which it's in the process of consuming. In his classic *Weeds,* the late Walter Conrad Muenscher, who was given to understatement, said that Japanese knotweed was "spreading rapidly and becoming obnoxious." Muenscher was too kind. I think it always was obnoxious.

For instance, cutting only prompts more growth, so that haphazard slashing results in more, not less, bamboo. And the thing has huge food storage capacities underground, probably in preparation for a nuclear attack, which I've considered.* Before I knew about its qualities, I chopped and mowed a plot to a grisly stubble one day. In two weeks the bamboo was chin high and laughing. It also sends out runners and rhizomes underground. In my neighborhood, it has leaped paved streets, which has caused one neighbor to hang cloves of garlic on his doors and windows. This is pure superstition because the plant also reproduces by seeds, against which garlic has no effect. A more practical solution is to fly over your yard at dawn in a fleet of gunships, blaring Wagner, and scorch the earth with lethal herbicides. But this isn't the kind of persona one likes to assume, even against weeds.

*A grove of real bamboo (almost as tough as knotweed) at ground zero at Hiroshima survived the blast and sprouted within days.

Besides, if you make a mistake with anything that can kill bamboo, it will do to your yard what God did to Sodom and Gomorrah. And nobody promises that anything will, in fact, kill bamboo. They say it could, or it may. As in "repeated sprayings of the Japanese-bamboo foliage with [dicamba and 2,4-D] could bring it under control" (which is the considered opinion of the authors of *Weed Control in the Home Garden*). As to what 2,4-D *could* do to any laboratory rats (escaped "experimentals") that happened to be hiding out in the bamboo, that's a matter of some dispute. However, a wild-eyed Health, Education, and Welfare advisory committee (no doubt under the demonic influence of Ralph Nader) once recommended that the herbicide be banned. With chemicals, there's always the chance of laying waste to the entire yard, or neighborhood, and leaving only the bamboo still alive.

As bad as it is, bamboo alone wasn't able to turn me against the whole vegetable kingdom. It took the rest of my yard to do that. It started with the locusts. (I don't mean the insects, for which I now feel a good deal of affection, but trees, which are members of the legume family.) The seed pods of honey locusts are good to eat. But the seeds, roots, bark, and leaves of black locusts are poisonous. (My guidebooks disagree about the flowers.) It isn't so easy to tell a black locust from a honey locust. I can't even tell them apart when I look at the pictures in the guidebook, and when I try counting leaves in the "wild" I'm lost. The only way I can figure to find out what kind of trees I've got is to eat a few pods, which I'm sure is just what the locusts have in mind.

I also have mulberries, gentle, friendly, fruit-bearing trees, I thought. But the unripe mulberries contain hallucinogens. I'm past the age at which I'm willing to gather and smoke unripe mulberries, nor do I want my yard filled with people who are willing. But that's the least of my problems. I also have horse chestnuts, pokeweed, and ivy, all poisonous in one part or another. I have rhubarb, the leaves of which contain oxalic acid. And around the house are planted yew trees, mountain laurel, rhododendrons, and azaleas, all bad. As one of my field guides says of the rhododendron family: "The ornamental bushes surrounding our houses may be deadly strangers waiting to kill the unwary . . . many of these plants possess a deadly poison that was used by the Delaware Indians as a suicide potion."

Nice shrubs. They have a clever little adromedotoxin that stimulates heart nerves and then blocks them, "leading to death by heart failure." Even the good parts of my yard are death traps. If capital punishment ever becomes widespread again they could bring condemned felons out here for a day in the country. Instead of the electric chair they could do a wild-foods lunch. "Here, Humongus, try a little of this rhododendron salad. Some pokeweed? Mountain laurel tea?"

I've even read of plants communicating, by chemicals, about the attacks of predators, so that they can all get the poisons out there in the leaf tips before the bugs start chewing. Not only did I have deadly strangers for foundation plantings, but they were also talking behind my back.

When I realized that this—the bamboo (which, to its credit, is edible) attacking my yard, and my bushes attacking me—was the famed community of life that environmentalists and evolutionists were always going on about, and over which I myself had occasionally gotten misty-eyed, I threw out my Darwin and started reading *Weed Control in the Home Garden* and the Old Testament. The first gave me technique, the second moral support. I reread Genesis to find out what God had to say about homeowners having dominion over their yards. What He said unto them (Adam and Eve in my reading) finally explained to me what had happened to my yard. After Adam and Eve ate the apple, God invented weeds. Before the fact, He said, "Replenish the earth, and subdue it" (easy for Him to say). After, He said, "Thorns also and thistles shall it bring forth to thee." I interpret the Bible liberally and take the thorns and thistles to include *Polygonum cuspidatum*. He also predicted how hard it would be to get the thorns and thistles out of your yard, saying "In the sweat of thy face shalt thou eat bread." I suspect this is the origin of the expression "In your face!"

I saw the light. I shed my evolutionist plant sympathies. I began to look on mowing the lawn as a Biblical struggle. When I got out the old rotary mower and filled it with gasoline, I thought to myself, "Subdue the earth," or, for variety's sake, "Smite the Hittites." I even hired a young man who smote a lot harder than I did. He did to my yard what Joshua was wont to do to the lands of his enemies. He "smote all the country of the hills, and of the south, and of the vale, and of the springs . . . he left none remaining."

Except the bamboo. I wouldn't say it's still standing, it's more like kneeling, or sneaking around in runners and rhizomes. And it has been forced to share the ground with grass. But it's far from dead. This spring I may resort to chemicals, perhaps even Wagner. But I'm not sure. Rumor has it that one man in town, driven bamboo-mad, assaulted the stuff with a chemical so bad that it's legal only in Texas. There's fear in the community now for future generations. But the fear is only in the human population. His bamboo is doing fine.

The Sociobiology of Humor in Cats and Dogs

There are times in life when you have to speak up. People may heap calumny or contumely on you, depending on their vocabulary. They may even ignore you. But you can't ignore the call of conscience. I'm talking about the kind of moral imperative that made Rachel Carson write *Silent Spring* and the Kingston Trio sing about Charlie on "The MTA." Not that I put myself in the same moral class as the Kingston Trio, but I too have a message that I feel compelled to deliver, whatever the consequences: dogs are better than cats.

I don't mean that I like dogs better than cats. Nor do I mean that dogs are better for certain things, like feet warming, while cats are better for other things, like appearing in fashion advertisements. I'm not saying that dogs are better looking than cats, or smarter, or that they have more cartoons drawn of them. They're certainly not cleaner. However, dogs have a sense of humor. Cats don't. And I can prove it. That sense of humor is what makes dogs better—purely, absolutely, ontologically better.

The only reason I bring this up is that I think cats are causing the decline of civilization. Two prime examples of their effect on modern life are the disappearance of literacy and the appalling condition of modern bookstores, both di-

rectly attributable not to MTV but to cat books and calendars. If you look at these things you'll see that they're
full of pictures and cartoons. Dog books have words. Think
of what dogs have done for literature: *Lad: A Dog* and *The
Call of the Wild,* the tales and cartoons of Thurber, even
Snoopy, who, although a cartoon character, appears in a
strip in which the name Beethoven is occasionally mentioned. Ranged against these we have Garfield and his sort,
who monopolize all the good spots near the cash register.
What would Shakespeare say?

I also blame cats for the rise of narcissism. Cats are the
ultimate narcissists. You can tell this because of all the
time they spend on personal grooming. Dogs aren't like this.
A dog's idea of personal grooming is to roll in a dead fish.
Dogs spend their time thinking about doing good deeds for
their masters, or sleeping.

Implicit in my criticism of cats is what might be called
the "petogenic" theory of culture. We've known for a long
time that individual persons tend to resemble their pets. I
think the same is true of societies. Do we want to become
a nation of Morrises? Is America to be nothing more than
a picky eater? Even physiologically cats are finicky; they're
obligate carnivores, or strict meat-eaters. Dogs, on the other
hand (or mouth), are omnivores in the true sense of the
word. Waste not, want not is the motto of all dogs. They
live in a permanent Third World of the mind where the
notion of throwing out food is incomprehensible. Let even
a King Charles spaniel off its leash and it will head straight
for the neighbor's garbage can, where it will consume anything that's edible, or has been next to something edible,
and spread the rest around the street. I think this is done
for moral and comic purposes, as well as those of appetite.
While cats are lying on divans waiting for the next flavor
surprise, dogs are reproaching us for throwing out perfectly
good gristle.

I didn't make this up about dogs having a sense of humor. Well, actually I did make it up. Then I found that some famous dead scientists agreed with me, like Charles Darwin. In *The Descent of Man* he said, "Dogs show what may be fairly called a sense of humour." In this opinion he was supported by George Romanes, another great man of science, also dead, in whose work I found the Darwin quotation. Romanes was one of the founders of the study of animal behavior and he credited dogs, but not cats, with "the emotion of the ludicrous."

What Romanes did credit the cat with was "disadvantages of temperament." I think he meant that cats aren't nice. There are those who say that when cats torture crippled mice (which they themselves have crippled) they're not enjoying themselves. You might as well say that William F. Buckley Jr. doesn't enjoy winning arguments. As Romanes says, "The feelings that prompt a cat to torture a captured mouse can only, I think, be assigned to the category to which by common consent they are ascribed—delight in torturing for torture's sake." Needless to say, neither Romanes nor I believe that laughing at a crippled mouse counts as having a sense of humor.

In contrast, Romanes said that dogs had the ability both to tell and appreciate a good joke. As one bit of evidence he discussed his own terrier, which "used, when in good humour, to perform several tricks, which I know to have been self-taught, and which clearly had the object of exciting laughter. For instance, while lying on his side and violently grinning, he would hold one leg in his mouth." So much for the intelligence of dogs. But I never said they were smart. And as jokes go, this isn't a bad one. Surely you've seen more than one comedian who would have been more amusing if he had put his leg in his mouth and grinned.

The usual criticism of Romanes is that he was guilty of

anthropomorphism. But as far as I'm concerned, animals *are* like people, and I believe science is coming around to this point of view. Consider the talking apes, the chimpanzees Washoe, Lana, and Nim, and the gorilla Koko. Not only does Koko have language, or something like it, but she also had a pet—a pet kitten. If having a pet doesn't make you like a human being, I don't know what does. (Too bad it was a kitten.) Furthermore, Donald Griffin at Rockefeller University has made persuasive arguments for animals having consciousness, at least when they're awake. And at the extreme left on the anthropomorphism issue, the animal rights movement claims for animals the same rights that human beings enjoy. The message of this movement is not that animals are *like* people but that they *are* people.

This new, and welcome, anthropomorphism doesn't quite prove Romanes's argument, however, and I did promise proof. For that we need sociobiology. If you recall, sociobiology is the scientific discipline in which you pick a behavior, like being rich, and construct a good story of how it evolved to show that it's genetic, and not the result of trust funds. (Critics sometimes claim that sociobiologists miss the subtle distinction between "inheritance" and "heredity.") If I tried this approach with my argument, I'd be in trouble. I can't start with my conclusion—dog humor—and then construct a story. That would be begging the question. So I'll start with the story. If it's good enough, I'm sure we'll agree that the behavior must exist.

This is my story:

Once upon a time, the ancestors of dogs—wolves in layman's language—found the stresses and strains of life in a hierarchical social species unbearable. They lived in packs dominated by alpha (not Alpo) males who pushed around all the other males. Everybody knew everybody else's business. Nobody had any privacy. And on top of that they mated for life. Consequently, they did the only thing possible: they developed a sense of humor.

How this worked in terms of evolutionary genetics is that the dog ancestors that had no sense of humor didn't live long enough to produce offspring. Put yourself in the proto-dog's position. Let's say you're a young male wolf without a mate. The rest of the males are always coming out of the den in the morning with big grins on their faces. By the time you get your turn at the moose carcass, all that's left are the chewy parts. You can't laugh it off, because you're genetically incapable of humor, so you lose control and attack the alpha wolf. He kills you. You know what that means—no offspring for you. Natural selection has just eliminated your no-sense-of-humor genes from the wolf gene pool.

It has, by the way, been scientifically demonstrated that wolves have a lot to laugh off. In a study of the wolves on Isle Royale, Michigan, in Lake Superior, the wolves succeeded in killing the moose only six times in 131 moose hunts. If these wolves were a football team they would be the New Orleans Saints. With ancestors like that, you end up appreciating the absurdity of life, not to mention moose hunts. And when you (a dog descendant) are faced with a human owner wielding a rolled-up newspaper, pointing to the rug, and shouting "Did you do that?" you do what your ancestors did when they missed the moose. You stuff your leg in your mouth and grin.

Cats are different. Cats have pride and dignity. Animals that live in hierarchical social systems (people, for instance) can afford pride and dignity only if they're at the top of the ladder. If you're a cat, you *are* the ladder. Put yourself in the proto-cat's place. Odds are you're a solitary carnivore. You live alone. Once in a great while you mate with a stranger who also has claws. On an average day you get up, kill something, eat it, and go back to sleep. What's funny? Laughing has no evolutionary benefits, and consequently you have no sense of humor. This is the evolutionary history of domestic cats. This is why cats are remote, independent, and mean. They don't tell jokes, and when you tell one, they don't laugh.

In other words, dogs are better (Q.E.D.). And they're more suited to us as a species. If you "miss the moose" at work a dog can sympathize with your plight. If you like, you can even blame the dog. It's genetically prepared to take this in good humor, and it will urge you to do the same when unjust accusations are leveled at you. In the same situation a cat will scorn you for being weak, and point out that when it feels bad, it kills something. I don't think this attitude is helpful. In fact, I firmly believe that as a people (or perhaps I should say "as people") we're less likely to cause a nuclear war if we keep dogs, not cats, as pets.

One final note: I know that emotions run high on this issue. It's conceivable that I might have offended some cat lovers, and that they might want to send me insulting letters, to which I would have to invent clever replies. To save them and me a lot of time and trouble, I thought I'd print the reply I have ready for them right now. It's a quotation from Romanes. After describing the cruelty of cats he said something that I like so much I'm adopting it as a response to all letters about this piece, perhaps to all letters about anything: "With regard to cats it is needless to dwell further upon facts so universally known."

Royal Flush: Travels in the Toilet Trade

I suppose the first order of business is to explain why I'm writing about toilets. I'm not doing it just so I can get in a little bathroom humor. Not that I'm opposed to low comedy, vulgarity, or shameless chasing after laughs. In principle, I'm in favor of all those things. It's just that my primary interest in toilets is in the mechanisms in the tanks, the valves and arms and floats, the things that hiss, roar, and gurgle, the stuff that is to the overall toilet what a central processing unit is to a computer. Call it toilet science.

I'm not claiming that it takes quantum theory or recombinant DNA to develop a good toilet, although it does take good toilets to develop quantum theory. (The modern theoretical physicist depends on indoor plumbing.) Inventing and perfecting toilet mechanisms does, however, depend on a knowledge of mechanical engineering. You need to understand water pressure and valves, too. I suppose that as an intellectual adventure, making a toilet tank valve may not compare to inventing general relativity or finding a cure for cancer. But if you consider the contribution of plumbing to human life, the other sciences fade into insignificance. Good toilets have done more for public health than all the doctors since Hippocrates. In fact, toilets are one of the few

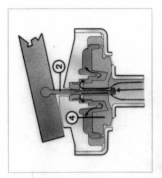

Float arm ①

② Fluted rod

④ Rubber disc

③ Inlet tube

Movable float

Outlet to tank

FLUIDMASTER

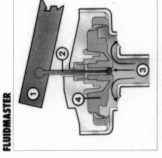

FILLING When the toilet is flushed the float arm (1) falls, pulling the fluted rod (2) up. Water (arrows) under pressure shoots up the inlet tube (3). The water pushes the rubber disc (4) up and squeezes between the disc and the lip of the inlet tube. From there it flows into the tank.

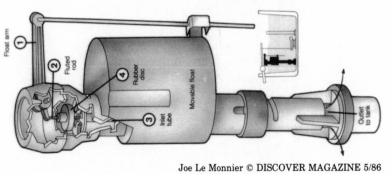

FILLED As the tank fills, the float rises and the fluted rod (2) drops. Now the hole in the rubber disc (4) is open, allowing water to flow under pressure into the space above the disc. The water pushes the disc down, sealing the inlet tube and stopping the flow of water into the tank.

Joe Le Monnier © DISCOVER MAGAZINE 5/86

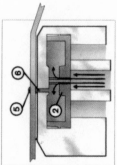

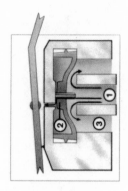

FILLING When the toilet is flushed, water (arrows) surges up the inlet tube (1). A small amount squirts through a hole in a rubber disc (2). The rest is deflected down by the disc through a system of baffles (3) and into the toilet tank.

FILLED As the tank fills, water pushes down on the diaphragm (4, in the large drawing). A rubber pad (5) on the diaphragm's arm seals a hole (6). This traps water above the disc (2), forcing it down against the inlet tube, and sealing the tube off.

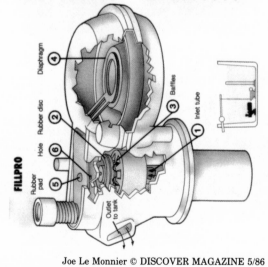

FILLPRO

Rubber pad
Hole
Rubber disc
Diaphragm

(5)
(6)
(2)
(4)
(3)
(1)

Baffles
Inlet tube
Outlet to tank

Joe Le Monnier © DISCOVER MAGAZINE 5/86

purely benevolent technological objects. Physicists may have known sin when they invented the bomb, but the worst thing the people who design toilet machinery are guilty of is a fondness for bad jokes.*

What attracted me to toilet technology is that it seemed comprehensible in a way that most of the world is not. I do not, in any fundamental way, understand my computer, my tape deck, or the telephone. I know that if I push the right numbers on the telephone I'll get my editor, or an ear-piercing whine that I figure is AT&T's revenge for being dismantled; that if I tap the right keys on my computer, this sentence will appear on the screen; that if I push the "play" button on my tape deck Johnny Cash will sing "I've been flushed from the bathroom of your heart."† But don't ask me why.

Toilets are different. The principle is simple. Water fills a tank. You flush. Some kind of valve lets enough water back in to fill the tank. In the old ballcock system, a float ball at the end of a metal arm rises with the water level, forcing down a stopper of some kind at the other end of the arm. Ballcocks are comprehensible, but lacking in romance. However, while all the science and technology buffs have been tinkering with computers, a world of toilet machinery has appeared. These mechanisms started to come out in the 1960s and '70s, and now they've taken over the hardware stores. They're plastic. They don't have long arms and float

---

*One company gives away ceramic coffee mugs shaped like toilet bowls, each graced with the legend "Think Tank." Obviously people in the toilet business have a sense of humor, although, given their sense of humor, it's a good thing they're in the toilet business.

†This song, by Jack Clement, also has lines like, "in the garbage disposal of your mind, I've been ground up dear." It's proof that C. P. Snow was wrong when he lamented the separation of the arts and the sciences in his famous essay "The Two Cultures." Look what happened when music (an art) met plumbing (a science). I plan to refute Snow soon in my own essay, entitled "Two Cultures Are Better Than None."

balls. Their inner workings are hidden. And yet they shut off water flow quickly, without that drawn-out hiss the ball-cocks are so good at. The new gadgets have the mystery of high technology, but they're simple enough to be understood completely, from flush to full.

Or so I told myself. The new mechanisms presented a challenge—to go where no science writer had gone before, into the toilet tank, and to penetrate what the 1611 charter of England's Worshipful Company of Plumbers called "the art and the mystery." I accepted the challenge.

I'd found in the course of my research that California was a hotbed of toilet tank valve development and production. Fluidmaster, the McDonald's of the business, in one competitor's words, was near Los Angeles, in Anaheim. Fillpro was two hours south in Carlsbad, just north of San Diego. Fillpro makes a valve like no other I've seen. It has no float at all and looks like nothing so much as a tiny, covered frying pan. Two small companies, both engaged in the search for the perfect flush, in California—it added up to only one thing: Toilet Valley. I flew to the Coast.

On the airplane I read a British history of bathrooms called *Clean and Decent* by Lawrence Wright. I recommend it highly. It's worth reading for the headings alone, among my favorites being "No Soap for Bacon," "Clean Queens," and "The Dauphin Commits a Nuisance." The book notes that the Minoans were among the first to have good plumbing. In the palace at Knossos they had terrific toilets with wooden seats, some of which may have had flushing mechanisms. The ancient Egyptians equipped their privies with stone seats. (It's a good thing they didn't live in Vermont.) Anyone interested in writing an unscrupulous best-seller, or headlines for the *National Enquirer* (and there are more of us than you might think), cannot help but notice that the pharaohs sat on these toilet seats. Since the pharaohs were supposedly divine, it would be safe to call these old stone potties the "Toilets of the Gods."

The toilets of the rest of us began to get fancy in 1596, when the first valve closet was invented, although it wasn't until the late 1700s that others like it came into use. Flushing toilets, as we know them, began to appear in the late 1800s. Several kinds were invented. *Clean and Decent* doesn't even mention Thomas Crapper, who I had been led to believe was the father of the flush toilet. Apparently he was just one of its many uncles.

From a contemplation of this long and distinguished past, I stepped off my plane into what I assumed was the toilet future—Los Angeles. At 3 A.M. my time, before I went to sleep in a vast bed that the Holiday Inn had draped with the finest petro-fiber blankets, I tried, like any good journalist would, to get the top off the toilet tank. I was unsuccessful. It was screwed down tight with one large Phillips head screw. By some inexplicable oversight I had neglected to bring a Phillips head screwdriver with me. Whatever the secret of Holiday Inn toilets is, it's safe. I wonder, do people steal toilet tank fill valves from motels?

The next morning, on my way to Fluidmaster, I made another research stop at Disneyland. Mickey was there, Pluto, Snow White, and some dwarfs. I saw Fantasyland and Frontierland. The park doesn't have a Toiletland, per se, but it does have toilets. From my point of view, however, they were disappointing. Disneyland toilets, at least the ones I used, don't have individual tanks with fill valves in them. I took my Mickey Mouse and Goofy dolls and headed up State College Boulevard to Via Burton.

Fluidmaster is on Via Burton. Fluidmaster is to toilet tank fill valves what Sony is to Walkmen. It makes something like six million fill valves a year. While I was touring Fluidmaster, it quickly became clear that plumbing science was different from other kinds of science. For one thing there was the decor. On the wall of one office at Fluidmaster was a calendar with a photograph of a lovely woman in a red bikini posed by the sea with a Ridgid Model 2A Pipecutter. If I remember right, her name was Lois. On a more technical level, it was a treat to stand in Fluidmaster's highly automated factory under an overhead conveyor belt that was carrying, slowly, 40,000 new toilet tank fill valves. It would be futile to seek a metaphor to explain what it is like to have 40,000 toilet tank fill valves pass overhead. I leave it to the imagination.

The valve itself turned out to be one of those things, like marriage, that are simple in principle but complicated in practice (see diagram on page 71). And, after all, why shouldn't it be? Why should something that saves so many people from typhoid and toilet hiss be easy to understand? I say this after having had a Fluidmaster executive talk me through a flush cycle, just the way Sky King used to talk down those twelve-year-olds in Piper Cubs. I crashed.

At first, that is; at first all I could see were the parts— the channel through which water flowed, rubber discs, a pencil-thin, fluted metal rod (controlled by a float), which moved through the rubber discs, opening them or sealing them off, and other stuff. The principle involved, I was told, was that of turning water against itself. Once the tank was full, the water was forced into a place where it pressed down on one of the discs and sealed the whole thing off. I failed to see how any of this occurred.

In fact, as we got deeper and deeper into the Fluidmaster mechanism, the purpose of the fluting on the rod, and the flexing of the discs, I felt myself slipping away. I didn't actually lose consciousness, although I was tempted. I didn't manage to stay fully awake, either. At the moment of deepest puzzlement I had a waking dream of the kind that has been known to crystallize other great insights. It came to me that what I was looking at wasn't a mechanism, but a little boy in funny clothes . . . from Europe perhaps . . . maybe Alsace-Lorraine. Then I saw the wooden shoes. The kid wasn't Alsatian at all. He was from the Netherlands! Fluidmaster had created, out of rubber and metal, a little Dutch boy who would reliably stick his finger in the right place at the right time, and do for your bathroom and mine what the original did for Holland.

I don't want to seem too smug about this. But I must say that having understood how so complex and important a piece of technology worked, I did feel a surge of renewed confidence, a certain legitimacy about living in the modern world, and a tremendous sense of relief. I was certain there was an aura of knowledge and power surrounding me as I drove down the Pacific Coast Highway, although that may have emanated not from me but from the brand-new flamingo red Camaro the rental car company had forced me to accept in lieu of my usual writerly sedan.

The next morning, after I stayed at an inn with a run-on toilet (which I fixed handily), I continued on to Fillpro, which is in Carlsbad in a new corporate park. It didn't look like a plumbing company. It had tinted windows and a color coordinated waiting room, which made me suspicious. Was Fillpro serious about toilets, or was it trying to get into computers? I'm happy to say that, judging from our extended, unbelievably detailed fill valve discussions, the company is very serious about toilets. And its valve is not only an engineering but also an aesthetic triumph (see diagram on page 72), if one can speak in these terms of toilet valves. It has no float, and no fluted rod. It has just one moving part. The whole thing sticks up only about three inches from the bottom of the tank. It too forces water to work against itself, but it does it *invisibly*.

I'm not going to tell you exactly how it works. It's a cardinal principle of science writing never to make your readers lose consciousness. But I'll give you a hint. The moving part may look like a frying pan, but it acts like a seesaw. When the pan end of the seesaw is down, that makes the incoming water do exactly what it does in the Fluidmaster set-up—shut itself off. I saw the valve at work in a see-through toilet tank, an object that I covet for my own home. I watched it flush and fill. No motion was visible in the Fillpro valve. It remained serene as the water rushed out and the water rushed in. Then all was quiet.

I don't yet have a transparent toilet tank in my home. I do have a Fillpro in one toilet, a Fluidmaster in another, and an old ballcock in a third. I think this is what journalistic objectivity is all about. But as inspiring as it was to learn how toilets work, I have to say that it wasn't the high point of my trip. That, as it turned out, had to do with the economic, not the scientific, side of plumbing. When I got to Fillpro and asked if I could use the bathroom, I was refused. The plumbing was out. During my entire tour of the plant the bathrooms were unusable, and plumbers were either searching for the problem or on their break. You can imagine how I felt. The prospect of a plumbing company about to be charged a fortune by plumbers will cause extreme and irresistible glee to toilet owners everywhere. I've nothing against Fillpro, but it's nice to know that while there may not be justice in the world, at least there's irony.

À la Recherche
des Crayolas Perdus

F rom the inexhaustible engine of commerce comes now
Aromance, the Aroma Disc system. Actually it came
a few years ago, but I've been trying to ignore it. It
doesn't mind. It sits patiently on the counter at my local
Captain Video, daring me to try it. (I guess selling aroma
records and videotapes in the same store is the retail equivalent of synaesthesia.) Aroma discs are about the size of
small computer floppy discs. You slip them into a box (only
$14.95), and as the disc is warmed, a mist or smoke emerges
with the scent of Passion, Fireplace, or After Dinner Mints.
I've resisted buying an aroma disc machine. The thing
frightens me. If an electrical fire is about to burn down the
house or someone left the gas on, I don't want Country
Moods confusing my smelling apparatus. More important,
I don't want to be part of a world in which a man and a
woman, on a romantic anniversary, turn to each other and
whisper, "Honey, they're playing our smell."

I'm sure people will say this anyway. My buying habits
seem to have about the same effect as my vote. And smells
are, and have been for a while, a big, strange business. On
a mundane level there's Lemon Fresh Joy and pine-scented
cleaners. Manufacturers stick "clean" scents in their laundry detergents. Scent-impregnated magazine inserts are now

so common that my three-year-old daughter thinks the only purpose of *Vogue* is to bring her perfume. She rubs them on her wrists—Opium and Obsession, which I suppose will be followed by Heroin and Paranoia—and asks me if she smells pretty.

Some enterprising souls are engaged in smell therapy, not helping people smell, or smell better, but using odors to help them feel better. Spiced apple is supposed to calm you down. International Flavors & Fragrances, the Dow Chemical of commercial smells, makes aerosol cans filled with the aromas of pizza, ham, new cars, anything that might entice somebody to buy something they might otherwise not. Society is everywhere losing its odor integrity. A Florida marine museum has sprayed its exhibits with sea smell. A few years ago Monsanto reported that it had synthesized a "fresh air" smell, which, of course, would be a silly thing to do if we actually had fresh air. And about three years ago, in what may be, aesthetically, the odor crime of the century, a company in Ohio came out with a cherry-scented garden hose.

It may seem that I'm some kind of weird odor curmudgeon, looking for something new to complain about now that television is all used up. After all, why not douse everything in a scent? Perfumes have been around forever. Human beings have always covered up one scent with another—as they should, sometimes. The paper diaper companies can perfume their products without criticism from me. I believe everyone who rides the subway should use deodorant. I'm all in favor of Chanel No. 5. But none of these uses of scents is deceptive. I know what's in diapers; I don't need to know what everybody on the subway smells like; and men and women who adorn themselves with perfumes aren't trying to pass as flowers. They just want to please their dancing partners.

It's the fake smells I don't like, the ones that are meant

to fool you. This is a dangerous business, because the human nose is fragile, emotional, and neither very bright nor, usually, very well trained.* Most of us have a poor vocabulary for odors. (The study of odors, however, has wonderful words, like "phantasmia" for an olfactory hallucination, and "olfactorium," the smelling equivalent of soundproof room, also called, in Latin, a *camera inodorata*.) Inside the brain, smell seems closely connected to the centers for emotion and cooking. It's snuggled right up to sex, anger, and blackened redfish. Obviously, the sense of smell is a prime target for manipulation. As perfumers have always known, love may be blind, but it's not anosmic.

Writers as well as perfumers know the power of smell. Marcel Proust set off the connections between smell (or taste, which is inextricably intertwined with smell), emotion, and memory with a crumb of petite madeleine swamped by tea. This doesn't work for me, perhaps because I don't know exactly what a petite madeleine is. But I recently discovered another substance that has roughly the same effect. What I use isn't a little French cake, but it is an object of childhood, as was Proust's madeleine. I stimulate my odor memory with a fresh box of Crayola crayons.

You see why I've yet to write a great French novel. But writers and smellers both have to work with what they have, and in our house we have children and crayons. I don't expect you to experience the Crayola smell just by thinking about crayons, since most people can't recall smells the way they can recall pictures or sounds. I expect you to go out and buy a box. Take it to a quiet (and not too smelly) place, an olfactorium if you've got one, flip open the lid, and

---

*Perfumers and wine tasters do train their sense of smell. Not long ago there was a man who sniffed fish for the U.S. Food and Drug Administration, and there may still be one. There being no acceptable test for rottenness in fish, for national fish security the government was relying on a human nose.

sniff. I don't mean wave it under your nose as if you were a courtier appreciating a jonquil. Get your nose, big or little, right down on the crayons, and inhale deeply. Pull that crayon smell right up into the old reptile brain. Once you get a good whiff of waxy crayon odor, the bells of child-fjhood will ring. Unless your parents beat you with those little sticks of "red-violet" and "yellow-green," you'll be flooded with a new-crayon, clean-piece-of-paper, untouched-coloring-book feeling—you're young, the world is new, the next thing you know your parents may bring home a puppy.

For a while after I discovered this effect, I did a lot of surreptitious crayon sniffing. Not that I was ashamed of it, but when my kids caught me, they took the crayons away. Children live by rigid social rules, and they know that crayons belong to children as absolutely as bourbon and scissors belong to adults. Up in my office, with the door closed, I would open the plastic bag from Toys "Я" Us (slowly, so it didn't crinkle too loudly), take out my box of 64 colors, and snarf up the crayon vapors. It took me some time, and some outside research, to get the higher levels of my brain involved so that I could see (not smell, you understand, but see) why it was that smelling crayons made me feel good. I finally realized that the smell of crayons isn't just an odor, it's part of our culture, something in the same class as the Howdy Doody song, and with the same resonance. In the future, long after they've stopped drawing with crayons, my daughters will have in their brains, as I do now, the useless and thoroughly inappropriate information that if you smell stearic acid you're about to have a good time.*

---

*That's the major component in the smell of Crayola crayons, according to the Crayola company, officially called Binney & Smith, which makes two billion crayons a year. That means there's at least one Crayola crayon for every family on earth. Not every family gets its crayons, however. In our house alone we've massed the allotments of whole crayonless villages.

If crayons can have that kind of effect, you can see why I'm wary about odor manipulation. Not the perfumes. The world can live with seduction. What I don't want is Ronald Reagan dousing my neighborhood with spiced apple smell to make me think I'm happy. I don't want the planet crop-dusted with "fresh air" smell, so I can't tell what I'm breathing. I don't want the smell of pizza to come out of a machine. I want to keep my nose tuned to reality, so I'll be able to smell a rat, or the blood of an Englishman. And I want everyone else to know it when I raise a stink because there's something fishy going on.

I like crayons because they have odor integrity. The Crayola people didn't stick stearic acid into their product to make you buy it. Nobody in his right mind would buy something because it smelled like a fatty acid. If there were a national odor museum, and there should be one (between the national restaurant and the national wine bar), I would give crayons pride of place in it.

And I would surround them with other objects with honest aromas that make up American odor culture. I have a few ideas of what these other objects should be. I got my ideas from William Cain of Yale and the John B. Pierce Foundation Laboratory in New Haven. Cain, who's also president of the New York Academy of Sciences, is involved in studying many aspects of olfaction. Other people might describe his line of work as the psychology and psychophysics of sensation. He calls it the smell game. As part of one of his experiments he had people sniff eighty everyday things, and he ranked the substances in terms of how recognizable their odors were. His list is the place for the aroma preservationist to begin, being full of things with wonderful and memorable smells.

On it are Juicy Fruit and Vick's Vapo Rub (remember getting it rubbed into your chest?), Ivory soap, baby powder, bleach, and pencil shavings. (Remember grammar school?

Remember driving that freshly sharpened pencil into the top of Billy Donnelly's head and telling him he would die of lead poisoning? Sure you do.) Cain also tested Band-Aids, nail polish remover, shoe polish (which reminds me of church), and Lysol. He tried coffee, chocolate, tuna fish, leather, mothballs, ammonia, and cigarette butts. Also bubble gum and bourbon. And, we can all be thankful, he didn't forget maple syrup, oregano, and barbecue sauce.

Crayons are on the list. They ranked eighteenth in recognizability. Coffee was first, peanut butter second, Vick's third. Not on the list, but favorites of mine, which I would lobby for, are rubber cement (which I remember from my newspaper days), newspapers (which I remember from my paperboy days), Cutter's insect repellent, and the unique, ineffable, and memorable odor of a bar I used to frequent in New Britain, Connecticut.

I know there will be some judgment calls here. Some people will want to preserve Brut aftershave and Herbal Essence shampoo, numbers 35 and 53 on Cain's list, and I will not. Some people won't want to have fresh cow manure in the museum. I think it's a must. Some people may even want to have all the smells on aroma discs instead of having the objects themselves. Those people will not be allowed to vote. Whatever the problems are, we should start solving them now; it's time to start paying attention to our odor culture. We're responsible for what posterity will smell, and like to smell. You tend to favor the odors you grow up with. As Cain points out, why else would anyone buy Noxzema? In other words, if we're not careful, we may survive the arms race only to end up alive but befuddled in a country in which everyone thinks garden hoses are supposed to smell like cherries.

Flea-Bite Economics

There are, in the fur of America's cats and dogs, four and a half billion fleas.* They live in relative obscurity. When people think of fleas at all it's usually in terms of circuses or disease. And it's true, there was that business of the black death. Rat fleas did help kill off a quarter of the human population of Europe. But they didn't mean to. They visited the plague on our houses because they bore, unwittingly, the bacteria that cause bubonic plague. And it might be remembered, as one flea specialist has pointed out, that the black death was no party for the fleas either. As goes the host, so goes the ectoparasite. A lot of fleas died too.

Things have changed since the fourteenth century. Bu-

---

*Or else I miss my guess. To get this number, which would be most accurate for September, after the fleas have had all summer to multiply, *I* multiplied the average number of fleas on the average infested dog—50 to 75 (say 50)—by the number of pet cats and dogs in the country, 56 million dogs and 34 million cats, or a total of 90 million flea bearers, according to the Humane Society of the United States. The result is 4.5 billion fleas. Of course, not all 90 million animals have fleas. But some have more than 50. My veterinarian has seen a small cat with hundreds. And there are also the uncounted strays and feral cats and dogs, which support a vast, uncounted flea population. So, as we say in statistics, it averages out.

bonic plague isn't what it used to be, partly because one doesn't encounter the rat flea any more in polite company. One doesn't *have* fleas oneself at all (unless one goes to a flea ranch, as I did, and carries some of the livestock away with one, as I also did). One is primarily concerned with the cat flea, which shows little allegiance to its eponym and lives, willy-nilly, on cats and dogs alike, and will also have lunch on a human being when it gets the chance. This is the flea of which there are, in the U.S., four and a half billion. And this flea isn't a major health problem for human beings.

I know that this is about as faint as praise gets. "Doesn't cause bubonic plague" isn't the kind of thing you want on a letter of recommendation. But these are insects we're talking about, and the cat flea has other qualities. It has never received the recognition it deserves for its contribution to the economy. In the face of declining heavy industry in America and a trade deficit with Japan, the cat flea has stepped in to stimulate commerce, provide jobs, and support an entire new industry devoted to its destruction. Nor does it hamper productivity or destroy crops. It just bothers pets. One can even assign a dollar-and-cents value to a cat flea— well, a cents value anyway. A virgin adult goes for anywhere from a nickel to six cents on the research market. That would make the four and a half billion worth $225 million—if they were all virgins. I suppose if they were all for sale at once that would drop the price down—probably way down—but the cat flea should be proud to be worth anything.

There are about 1,600 species of flea, although most are of no economic importance. Nonetheless they do seem to attract money. Miriam Rothschild, of the British Rothschilds, has made a career of studying fleas. She has catalogued the Rothschild flea collection at the British Museum (Natural History). She has done considerable flea research,

and in 1965 wrote in *Scientific American* both wittily and
knowingly of fleas. She described, for example, the cat flea
responding to exhalations of warm carbon dioxide to find a
host, and another flea, which specializes in the large jird,
"a rodent that lives in the sandy soil along the Ili River of
central Asia." The jird flea can tell when a man walks by,
and will "emerge and pursue him for quite a distance." A
frightening thought. Rothschild also described what one
might call the personality of fleas in one of the gentlest
paragraphs ever written about something that has six legs
and sucks blood. "Parasites," she wrote, "must be modest
in their demands and unobtrusive in their ways."

True to her description, fleas don't put great demands on
a host. Dogs and cats that don't develop allergies to flea
saliva can support a population of fleas with relatively little
discomfort. This, of course, is the opinion of human beings,
who are notorious for underestimating the distress of others. Modest or not, fleas are determined. The cat flea may
not walk a mile for a camel, but it will hop thirty feet for
a Labrador retriever. Once it gets on the retriever it will
breed at a good rate. If you start with ten female fleas, in
a month you could end up with 100,000 eggs (with close to
100 percent hatchability) and 1,800 adult fleas, assuming,
of course, that you've got twenty or so dogs to feed them.
And these fleas wouldn't be that easy to kill. If you take a
flea in your hand and try to squash it between thumb and
forefinger, you won't succeed. Fleas are too well armored.
You've got to get them on a hard surface and crush them
with a fingernail, a knife, or, if you're in the mood, a ball-
peen hammer.

Or you can use a flea spray. I don't know who invented
insecticides, but some of the credit should go to the insects.
And insecticides, in this case flea-icides, are important to
the economy. This is the basis of flea-bite economics. I know
it's hard to accept my idea. You have to change your notion

of fleas, or of economics. And it's possible that there's some horrible flaw in my reasoning, but I can't find it. It seems to me that cat fleas, much like the individuals in *The Wealth of Nations,* promote the general welfare by pursuing their own bloody self-interest. Like this:

Pets get fleas. Pet owners buy flea shampoo, flea collars, flea sprays, and foggers, or "bombs," to purge afflicted houses. All of this means money pumped into the economy. To be sure, from the pet owner's point of view the money could have been better spent on hamburgers or X-rated videotapes. But from the point of view of the economy, money is money. And in this case it's $600 million a year worth of money. That's how much consumers spend on "ectoparasiticides" for the "companion animal market"—in other words, stuff to kill fleas and ticks on cats and dogs.

I myself have seen the innards of this companion animal flea market. I went, not wearing any kind of protective collar, to a flea ranch in Dallas. I call it a flea ranch. It calls itself the Zoecon corporation. It makes Vet-kem and Zodiac flea collars and sprays (sold through veterinarians and pet stores), and it was a Zoecon executive, who works in a shiny new granite and glass office building, who gave me the dollar figures on the flea economy. They don't keep the fleas in that building. They're on the other side of a parking lot, in a smaller building. But, then, fleas are small.

Zoecon has a small part of the flea business, but it's at the cutting edge of flea science. Carl Djerassi, a pioneer in the development of the birth control pill, was one of the founders of Zoecon, which has come up with something similar for fleas, although that product is more in the line of life control. The patented substance is a synthetic hormone called methoprene. Given in large doses at the wrong time (for the flea), it interrupts growth so that flea larvae, instead of wrapping themselves in cocoons, like butterflies, and then emerging as adults, just never grow up. They kind of fade away. The nice thing about methoprene is that it doesn't do the same thing to human beings.

When I visited Zoecon, about 250 people, 65 dogs, 70 cats, and 100,000 fleas were working there. The animals are involved in product testing and flea rearing.* The flea wrangler puts a herd of adult fleas on each of six cats that, at any one time, are working as flea food. The fleas feed, mate, and lay eggs. The eggs are collected and hatched, and the larvae—teeny, bristly wormy things—are raised to adulthood on a diet of dried beef blood and Purina Cat Chow.

At the flea ranch I learned that if you want to anaesthetize a flea you use carbon dioxide. A little bit excites them, and a little bit more knocks them out. The reason for knocking fleas out is that they're hard to handle when they're awake. You can't just scoop up fifty fleas in a tablespoon and sprinkle them on a mongrel like jimmies on ice cream. But if you run a plastic tube from a big CO_2 tank to a jar full of fleas, and open the valve, the fleas turn into quiescent black dots, as docile as sand. Until they wake up. Then they tend to jump on visitors. The man at Zoecon who conducts the tests on dogs picked one runaway flea off my neck as it was about to drill through my skin. When I got back to my hotel, undressed, and shook out my clothes, another flea leaped out. And I had no CO_2. Naked and unarmed, except for a hotel issue plastic shoehorn, I hunted the insect down and killed it. The process took about fifteen minutes and would have done Chaplin proud.

I don't think I brought any fleas home with me, but since

---

*The latter, done on cats, isn't quite so bad as it sounds. Although it's not good either. Cats that develop an allergic reaction are taken off flea rearing for a rest, so the ones that are working aren't actually itching all the time. You could say that at least these cats have jobs. Not everybody can be a high-living Siamese. The cats come from the pound and if somebody hadn't hired them, their prospects would have been bleak. Also, they're doing their bit for cat-kind, although since they're cats, this probably doesn't mean much to them.

I went to Zoecon, I've been troubled by something else. I made up the flea-bite economy to be funny. But Zoecon is a real company where people make real money. What if flea-bite economics isn't a joke? (I'll bet this is what happened with the supply-side theory.) I've searched for a way to disprove it. You could say that the money spent on fleas is thrown away because nothing useful is produced. But then what about the money spent on movies? Are they useful? I suppose it depends on the movie. But even the worst are good for the economy. And what about dandruff shampoos? Is dandruff good for the economy, and bad breath, and body odor, not to mention cockroaches, ants, and silverfish? Are they all good too? I think so. It seems inescapable: everything that's bad, but not that bad, for which it's possible to market a remedy is good.

As disconcerting as this idea is—both economically and grammatically—it's also reassuring. It solves a long-standing problem in theology. (They claim that science doesn't have anything to say about religion, but they're wrong.) For many years, certain people have claimed that God put everything on earth for a reason. Other people, usually adolescents, have said, "Yeah, what about fleas and dandruff?" These are things that aren't bad enough to test our souls, and nobody nowadays believes that dandruff comes from the devil (fleas maybe, but not dandruff). So why are they here? The usual answer has been that God moves in mysterious ways (like a knight in chess), and we need not understand Him, but I think even the people who take this position have always been a little embarrassed by it. Now there's a better answer. These things are on earth for a reason, and it turns out to be the same reason that everything else is here: to help us make money. So, thank God for fleas.

Welcome
to Planet Photon

O Z and I—OZ is his code name, mine is GOR—are considerably older than most of the other Photon Warriors. We're standing by our base goal, listening to a disembodied woman's voice tell us about our mission. The voice is husky, and either menacing or seductive (I always get those two confused). Most of the time I can't understand what the voice is saying, but I don't think that's supposed to be part of the game.

With our phasers at the ready, OZ and I discuss strategy. We've been getting murdered. We keep getting an awful "Buzz!" in our helmets that means we've been shot (-10 points). Usually we don't even see who's shooting us. We run, hide, spin, shoot, and what we get is more buzzes. Other people, like CD DISK, are scoring 600 or 700. We've been getting scores like 140 and -90. Not this time. This time we have a plan. At the end of the countdown we sprint to the sniper's nest. We hunker down, picking off the red team players (10 points per hit) when they come out of hiding. Sometimes we get picked off too. As the game wanes, we rush their goal (200 points if you hit it three times in a row). OZ tears down a ramp into a corridor, flattens himself against a wall, jumps out into the open, spins, shoots, gets shot. I make my dash to a tunnel, where I start getting buzzes from an invisible enemy. I look for cover. Crouched in the dark, hiding, I ask myself: Is this any way for a grown man to spend Friday night? And what's my score?

Incredibly, our strategy doesn't pay off. After the game, on the big TV screen in the lobby, many lines below CD DISK, are GOR 180, and OZ 40. If you think I'm depressed, you should see OZ.

Back in the old days, before Photon, before you could get *inside* a video game, there was Asteroids, and Space Invaders, and what, for my money (and I spent enough on it), was, and still is, the greatest video game of them all—Missile Command. I can't say I was a great player. At the Times Square arcade where I spent my lunch breaks I was just a hacker. Of course, the competition was tough. Times Square is to video games what Times Square is to crime—and it's the same people who are good at both. However, once, on a trip to Brattleboro, Vermont, I played Missile Command at a little arcade, a spot of bright urban blight beeping and flashing in the midst of the health-food stores. I got the highest score the machine had ever seen, and left singing "You don't tug on a Superman's cape, you don't spit into the wind, you don't pull the mask off the old Lone Ranger, and you don't mess around with Jim." At least not in Brattleboro.

I was deeply involved in video games. I still think there's no music as sweet as the noise of the Playland arcade at 48th and Broadway with all the games going at once. It's like being in an electronic rain forest. And there's nothing like seeing that the top three scores on a game are 999,999 (as high as the counter goes), and that three players who beat the game are GEE, MAX, and GOD. When you see that you say to yourself, "This is the place."

I myself never noticed any negative effects of video games. Sometimes, after a three-hour, $15 Missile Command lunch I would feel cheap and used—not if I had a good score, though. I did stop playing, but only because I started working at home, far from the good arcades. Then, finally, I moved out of Manhattan, bought a house, and had children. I mowed the lawn. I learned how to use a screwdriver. And I took up tennis. Now that's a game that can really make you feel cheap and used.

So, when Planet Photon was born I was ripe for recruitment. Certain things were missing from my life, like sound effects. And I wasn't yet prepared to see myself as others (my neighbors) saw me: round-shouldered and beleaguered, pushing a double stroller with two kids in it, dragging along two untrained, ungroomed dogs in a circus parade of domestic confusion—in other words, as the exact opposite of Arnold Schwarzenegger. I wanted to see myself differently, as a guy in a helmet, with a phaser in his hand, zapping the Klingons. As soon as I heard about the game, I started practicing my moves.

Planet Photon was invented by George A. Carter III, the man who also invented, and I quote his news release, "the world's first motorized surfboard." Remember the motorized surfboard? No? Well, as Carter III himself has admitted, he "anticipated the market a little too soon." Ah, but then he thought of Photon, a cross between a science fiction movie and a very large video game. The promotional literature on the game mentions the movie *Tron,* a fantasy in which people enter a video game. According to Carter, with Photon, the fantasy is now reality.

Photon is played on an indoor field of 10,000 square feet. The basic field (three of the nine existing Photons have the more complex Omega field) is high-ceilinged, carpeted, and has two levels. There are ramps, tunnels, some tacky lighting, music, hidey holes, walls, base goals—pretty much anything you might want in a planet. There's also one slightly grotesque touch, a gallery, or walkway over the field, from which spectators may watch the game, or, for a dollar, use one of the gallery's phaser stations. From these stations you take target practice. You can shoot the players. But they can't shoot you.

The players must shoot each other. To this end, they're issued helmets, battery belts, chest pods, and phasers, and allowed to sign on with any code name they like, "Ace" or "Ed Norton" (who was very good) or "Fooltron" (who wasn't). There are referees to explain and enforce the rules. You're not allowed to stick your phaser through grates or around corners, run at ninety miles an hour down the ramps, or get within five feet of another player. (I'm sure full-contact Photon will evolve in the near future.) Unlike tennis, Photon has sound effects. Helmet speakers ring with zings, zaps, and buzzes when you hit or miss opponents, or get shot yourself.

Naturally, computers are involved. The scorekeeping and sound effects are managed by two IBM PCs. I suppose that, depending on your point of view and how good a shot you are, you might look on Photon as part of the humanization of technology or as just another form of electronic leprosy. I'd say that Photon is a positive contribution to human life, but not that positive. I would compare it in value to the introduction of the three-point shot in professional basketball. The difference, of course, is in the infrared light (this is what the phasers shoot, not laser beams) and the little computer chips in each player's chest pod, which register hits and communicate, by means of radio waves, with the PCs. Larry Bird, as we know, doesn't wear a chest pod, although I'm not entirely convinced that Kevin McHale isn't at least partly electronic.

I've now played Photon in Dallas, by myself, and in Kenilworth, New Jersey, with OZ. (The shooters in New Jersey are tougher, although I can't say why or Don Corleone will send somebody to break my legs.) And I'm sorry to say that

Photon is nothing at all like being in a video game. The two experiences are physically, and metaphysically, opposite. At its peak, video game play is an "out of body" experience. What happens is that after an hour or two the brain reaches a state of deep relaxation, a technologically induced trance. The mind enters the game. Like Zen archers who can hit the target blindfolded because they *are* the arrow (at least that's my idea of how they do it), you *are* the missile, or the ray, or the blip. The self disappears, along with the self's quarters. Action occurs without the usual physical limits. Maybe your fingers are limited by the speed of light, maybe not. Maybe they're your fingers, maybe not. Until you run out of change. That's when the self re-emerges and drags the fingers kicking and screaming out into the harsh and sobering light of Broadway.

Photon isn't like this. For one thing, it costs more than a quarter to play. It's $3 or more per six-and-a-half-minute game, depending on where you play, and $4.50 or more for the membership fee. More important, Photon is a decidedly in-body experience, and in my body this isn't such a great thing. If they had rented me Ivan Lendl's body (or Martina Navratilova's—what the hell, as long as it's a Czech) my scores might have been higher. But the game isn't that advanced yet. They only provide helmets and phasers. It's BYOB (bring your own body). And you never know, until you try it, what popping up and down from behind good cover and sprinting through tunnels in a modified duck walk will do to your thighs.

Thigh pain isn't the stuff of fantasy, at least not my fantasies. In Photon it's impossible to believe, for more than ninety seconds, that you're anybody other than yourself, and I can get that at home for free. It's not just the limits of your own body that serve as reminders of dull reality. There are other bodies in this game too. In particular, there are a lot of junior high school kids who sign in with names like Psychopath or Gaddafi (this kid wasn't well received in Dallas) or Lord Corwin (from a series of science fiction novels), and then wander around, chubby and befuddled, getting riddled with light beams by eighteen-year-old ne'er-do-wells who are presumably taking a break from substance abuse. Of course, the advantage to having little kids in the game is that everybody gets to shoot them. OZ, who teaches junior high school, found this particularly gratifying. And it was in the game with the kids that we both got our highest scores.

As to the effect of Photon on the kids themselves, I don't think it will rot their brains. Drugs and television will rot their brains. Infrared light is no big deal. Besides, Photon will keep them in good enough physical condition so that when they're my age they can play tennis. And I don't believe, as I'm sure somebody does, that Photon is part of the Ramboization of America. I suspect that pinging people with toy phasers is far less inducive to violence than watching one of Stallone's ketchup classics. The sad truth (for those of us who like the occasional evening on another planet) is that although there's a little shooting going on, the big difference between Photon and Capture the Flag is that Photon uses more electricity.

Nonetheless, the game seems to be a success. According to Carter III's office, 147 franchises have been sold, including the licensing of twenty playing fields in Japan. Everybody I've talked to agrees that the Japanese will love the game. Most people think it's the electronic gadgetry that will attract them. I think it's also the code names. A friend who just returned from Japan with a huge hangover told me that businessmen there, when they go drinking, use special names (not their own) so that they won't be ashamed of what they do and say. Such names are ready-made for

Photon. And once the game takes off in Japan I think we can rely on the Japanese to take it a step further and produce a street and subway version. In this game each player will carry a completely self-contained, miniaturized apparatus about the size of a portable tape player. It will be called, naturally, the Sony Hit Man.

In Vino Sanitas

I was so relieved to hear that wine is good for my health. You see I used to be a wine bore. And I never noticed that drinking did anything for my health, or my personality, although it did give me something to talk about. At the zenith of my interest in wine (which coincided with the nadir of my friends' interest in it) I could go on for minutes about the quality of the soil of Saint-Émilion and Graves, about the varieties of grape grown in the Médoc and Pomerol. I could do a complete exegesis of a German wine label. I knew what *Trockenbeerenauslese* meant, and I could say "noble rot" in three languages.* I could also talk the nonsense of the palate: I found berry in the Zinfandel, a trace of ash in the Cabernet, and, once, a suspicious hint of PCBs in an overly supple young Chardonnay from the state of Washington.

At some point, sadly, I felt I had said everything I had to say about tannin. My conversational level went straight downhill, which is to say in the direction of "drinkin' half gallons and callin' for more." This latter kind of wine talk

---

*German, *Edelfäule;* French, *pourriture noble;* and English, noble rot. *Trockenbeerenauslese* means, roughly, "selected shriveled berry picking." What that means is another story altogether.

was reported as early as the sixteenth century by that great wine writer, François Rabelais: "Page, my dear boy, fill this up till it spills over, if you please.—Wine red as a cardinal's hat! . . . Toss it off like a Breton!—Down in one gulp. That's the stuff." This account of an early wine tasting appeared in *Gargantua,* and Rabelais called it the Drunkards' Conversation. As it makes evident, without disquisitions on the aesthetics of the nose and palate, drinking wine becomes less like going to an art gallery, and more like plain old drinking. One has only the experience itself, and not the endless conversation about the experience. It hardly seems worth it.

That's why I'm so happy to see medical science enter the vineyards. I've been convinced for some time that the purpose of science is to provide something to talk about when art and poetry and music lose their appeal (and how often can you sing "Wine spo-dee-o-dee, drinkin' wine . . ."?). Now, what Masters and Johnson did for sex and Einstein did for trains, medicine has done for drinking. No longer when faced with the question of what to say about the 1980 Brane-Cantenac you're having with your carpaccio,* need you babble about why it matured earlier than the '79s, or is so much in the shadow of the massive '82s. Instead, you can raise the question of whether, as you drink it, it's increasing the level of high density lipoproteins in your blood stream to scour away the atherosclerotic plaque that the carpaccio is in the process of depositing. This is a whole new aspect of the complementarity of wine and food.

The origins of the current fascination, if not intoxication, with the idea that drinking is good for you lie in several epidemiological studies that show that people who have one

---

*Carpaccio is neither a character from *Romeo and Juliet* nor a friend of Frank Sinatra's, but raw beef sliced extremely thin and served with tasty sauces like tarragon mayonnaise.

or two drinks per day (of anything alcoholic) are likely to live longer and have fewer fatal heart attacks than people who drink a lot or people who don't drink at all. At first glance this seems like great news. There is, however, the problem of sticking to one or two drinks a day. Perhaps it takes a certain kind of person to approach alcohol in such a measured, unenthusiastic way, and it's being that person (not trying to mimic his habits) that defends against heart attacks. This person is not too tense, not too loose, not a stiff-necked teetotaler, but not someone who figures that as long as the bottle's open he might as well finish it. He's steady, unexcitable, predictable; he makes sure he gets his bran in the morning. He's able to treat a bottle of Le Montrachet (one of God's great gifts to drinkers) as if it were a dose of salts. He is, in short, a person in whom moderation has reached the level of irredeemable excess.

For those of us who aren't this person, who occasionally stray and drink three glasses in a given day, there are other good things about wine. I learned about them through the Wine Institute, an organization that works on behalf of the California wine industry. Some of these industry organizations, like the National Live Stock and Meat Board or the Tobacco Institute, are a bit biased, but I think the Wine Institute is pretty objective. Its attitude seems to be summed up by a headline in a booklet it publishes called "Wine and America." The headline reads WINE IS CIVILIZATION.

The institute recently sponsored a symposium on Wine, Health, and Society. (I didn't make it to the meeting because of a Gewürztraminer hangover, but the institute was kind enough to send me transcripts of some of the talks.) Several doctors appeared and spoke on behalf of wine, but I didn't take that too seriously. What impressed me was that Jane Brody appeared at the symposium and had good things to say about wine. Jane Brody, for those of you who haven't been terrified by her, writes the "Ways to Die" column in *The New York Times* (the newspaper calls it "Personal Health"). Each week she writes about some new thing that can kill you (or make you sick to your stomach) and how to avoid it. I never remember how to avoid it: I just remember that it's going to kill me. She also writes books about food, and I think it's fair to say that she has a more negative view of pastrami than any other literary figure. She's very critical of cholesterol and very much in favor of grains, the wholer the better. I know in my bones, and in my arteries, that she must hate carpaccio. I think of her as the Eighties' answer to Rabelais. He says yes. She says no. Except, oddly enough, to wine.

Now I don't want to give you the idea that she, or the other speakers, or the Wine Institute, advocates heavy drinking. They don't. They advocate moderation, except for one suspicious passage on vitamins in that "Wine and America" booklet. It begins, "Half a liter of wine (about $4\frac{1}{2}$ four-ounce glasses) supplies the following vitamins: 5 percent RDA (recommended daily allowance) of riboflavin, 2 percent RDA of niacin . . ." and it goes on to pyridoxine, folate, biotin, thiamine, and B_{12}. I worked this out on my calculator. To get your daily ration of riboflavin (forget the niacin) you've got to drink ten liters of Gallo Hearty Burgundy. You could, I suppose, get some of it from food, but when you're drinking that much it's hard to find time to eat.

In addition to providing vitamins, wine relieves tension, and even with the alcohol removed it helps you deal with electric shocks. In one study referred to by Brody, rats that were given "wine residue" (wine without the alcohol) showed reduced sensitivity to shocks, as did rats that guzzled real wine. Furthermore, moderate drinking (of any kind of alcohol) increases the levels of high density lipoproteins. We

don't know yet if these are the *good* high density lipoproteins, the ones that fight cholesterol, but I figure it's better
to be safe than sorry. Wine promotes absorption of nutrients
from food, and wine seems to show anti-viral activity. I also
believe it has a positive effect on wit. While driving through
New York's Bowery the other day (where the so-called winos
hang out) I was approached by a man who had clearly over-
indulged in some lesser vintage of Night Train or Thun-
derbird. He had a cardboard sign with the plea, "I owe
$250,000 in back rent, and I'm a quarter short."

Not only is wine good, it's better than hard liquor or beer.
John De Luca, president of the Wine Institute, has proposed
a health and safety index for alcoholic beverages to replace the old "alcohol equivalence" in which a beer equals
a shot equals a glass of wine. Wine, as he would have it, is
just not a boozer's drink. It's the "beverage of moderation."
Its alcohol is absorbed into the blood stream more slowly
than that of hard liquor. And it plays a different role in
society than other beverages that produce what one physician at the symposium referred to (condescendingly, I
thought) as "the effect of alcohol on the central nervous
system." De Luca mentions, for example, "wine's sacred
place in Catholic and Jewish religious ceremonies." A good
point. I can't see Miller Lite at a seder. And I know that
none of us would want to see John Paul II holding up a
frosted glass of pilsner in St. Peter's Basilica during the
transubstantiation.

Since wine is so good, I don't see how I can afford not to
drink. If I suffer from a little hangover, the viruses are
suffering too. And if I lose on heart attack prevention because I drink a bit too much, I nonetheless gain tremendously in reduced sensitivity to electric shock.

I didn't know about all these benefits a few years ago
when I was a member of a wine appreciation group. It wasn't
a big group, like the Society of Medical Friends of Wine or

Les Amis du Vin, or *Les Grands Chevaliers du Tastevin.* It consisted of myself and two friends and it was dissolved after two meetings for gross lack of moderation and failure to stick to the point during the discussions. But now that we have some new stuff to talk about I'd like to revive it, and rename it. In the past we called ourselves *Les Grands Amis des Chevaliers du Tastevin mit Prädikat.** I think it should now be called *Les Grands Amis des Chevalier du Tastevin mit High Density Lipoproteins.*

And the next time we meet, after we inhale the bouquet, check the wine's legs, swirl the stuff around in our mouths and swallow it, our conversation will be nothing like that of Rabelais's drunkards, but more along lines that the Wine Institute would approve of: "Could you pour a little more? I'm short on my niacin.—I like it. Very robust chromium content.—Hey, good silicon.—All right! Let's drink to magnesium absorption!—Here's to bioavailability!—The Framingham study!—Zinc!—You know what I like about this wine?—What?—The effect of its alcohol on my central nervous system, that's what. Pass the carpaccio."

*The Big Friends of the Knights of the Little Silver Cup with a Predicate.

Personology Today

I used to think that *Tropical Fish Hobbyist* was a great magazine. Before that I liked *Fly Fisherman, The New Yorker,* and the *Paris Review.* When I was even younger, there was *Field & Stream* and *Boys' Life,* which always had great articles on how to build things (tables, chairs, small buildings) with a jackknife. All that's behind me now. I've found a magazine that has it all, and more. It's got poetry, it's got drama, it gives you a glimpse of a strange world. It's even got good advertisements. It's called the *Journal of Personality* and I don't think anyone can afford not to subscribe, even if it does cost $27 for four issues.

I was led to the *Journal* because of a news report of some new findings (by psychologists) on love. I always like to see what happens when scientists step onto the poets' turf. In this case I found two love papers, both of them in prose, but with some nice turns of phrase, such as the description of genuine love as a "rare expression of optimal functioning . . ." and even better the reference to "Love items with rotated factor loadings." (I love it when they talk that way.)

In one of the articles there was also a sensible use of a quotation from Shakespeare. I can't tell you how rare this is. Scientists love to quote Shakespeare, but they almost always do it badly. The usual trick is for an ear, nose, and throat man to dig up a line like, "I'll speak in a monstrous little voice" (from *A Midsummer Night's Dream,* of course), and slap it on top of his latest paper on laryngitis. Well, David McClelland of Harvard firmly anchored his quote in the text of his paper for the *Journal,* and it was right on target: ". . . Love is not love/ Which alters when it alteration finds,/ Or bends with the remover to remove:/ Oh no! it is an ever-fixed mark,/ That looks on tempests and is never shaken" (Sonnet 116). McClelland was arguing, as was Shakespeare, for a view of love as characterized by joy, constancy, and altruism.

I admit that in and of themselves a couple of good phrases and some poetry aren't going to pull readers away from *Fly Fisherman.* And the *Journal* does have some real drawbacks. There are no photographs, which puts it at a disadvantage to more colorful publications. But, having looked through a number of issues, I can tell you that it does what all great magazines do: it expands your awareness. Four times a year it brings into your home (as magazine publishers love to say) the world of psychology, a world that is, in its own way, every bit as exotic as that of the angelfish or the largemouth bass.

The words psychologists use are themselves worth the price of a subscription. These are words you don't see anywhere else, like "personologist." I don't have enough room to quote the title of the article in which this word appears, but the paper is about how the non-psychologist, you or I, makes judgments about people he meets. Everybody does this of course, but in the transforming language of psychology, everybody becomes an "intuitive personologist."

This is a matter of style. The content is even more intriguing. I don't mean the conclusions the scientists come to, which aren't so interesting, but how they get to those conclusions, how psychologists work. I never knew that in a lot of these studies, students are required to be experimented on in order to get course credit. For instance, in a breakthrough study on fidgeting, which I'll discuss in detail later, "students participated in the study as part of a course requirement." In other cases students "volunteer," but get partial course credit for that invaluable part of any liberal arts education: being an experimental subject.

The way I read it, this means that a practicing academic psychologist can get tenure, write a textbook, and then teach a course in which students are required not only to buy his book but also to be subjects for whatever experiments he wants to do on them. Usually the kids just have to answer questionnaires. But not always. In a study at the University of Toronto on candy eating (published, I'm sorry to say, in a competing journal) subjects were "pre-loaded," which means that they had to drink not one, but two eight-ounce Borden milkshakes, one chocolate and one vanilla, *before* they started eating the candy. Is this any way to treat the daughters (all the subjects were female) of the middle class? Isn't this what convicts are for?

If the milkshake business seems gruesome, it's nothing to the work on fidgeting. Until recently, I was unaware of the large body of work on fidgeting (why people fidget, how people fidget, who fidgets). Fortunately, "An Analysis of Fidgeting and Associated Individual Differences" by Albert Mehrabian and Shari L. Friedman of UCLA, published in the June 1986 issue of the *Journal,* had a little review of some of the classic work. Jones, for instance, in 1943 "required subjects, male and female, to drink 10 glasses of water over a two-hour period while strongly discouraging their use of bathroom facilities during this period." You may wonder what happened. Surprisingly enough, "for subjects who did not use the bathroom facilities, there was a significant increase in nervous movements, particularly in the genital and leg areas, compared with subjects in the control group who simply rested for two hours." Yet other researchers, forgoing the water treatment, had a system in which schizophrenic patients earned pennies (of course this was in 1973, when a penny was a penny) on a "fixed interval schedule," which I take to mean that every so many minutes the sane person gave the crazy person a penny. Well, when the researchers started taking longer to give out the pennies, the poor schizophrenics showed "increased pacing and water consumption," as you can well imagine. Unfortunately, the researchers didn't think to capitalize on this serendipitous turn of events and take the study one step further by discouraging the schizophrenics' use of bathroom facilities. Too bad. When opportunities like this are missed in science, they seldom arise again.

For their work, Mehrabian and Friedman took a new tack, using neither water nor schizophrenics. In three separate studies, they got bunches of undergraduates together and gave them questionnaires about fidgeting. In one of the studies they also had someone play checkers with each subject and take a long time to make the moves while someone else watched to see if the subject fidgeted. (Of course, one never tells the subjects the truth about what's going on, for obvious reasons. If the schizophrenics knew what the penny thing was all about that would have influenced their actions, one way or another.) On the questionnaires were statements like "I scratch myself a lot; I usually jiggle my pen when I am holding it, but not when writing with it; I often bite my lip (on purpose); I hardly ever suck on my tongue." I could tell from the questions that I'm a fidgety person. I jiggle my pens, and chew them and my fingernails, and grind my teeth. However, I want to make one thing clear. I don't suck my tongue. That's disgusting. Sometimes

these kinds of studies don't come up with firm or comprehensible findings. But the last sentence of this paper is
pretty clear. It says the studies led to "the conclusion that
fidgety persons were either of an anxious or a hostile temperament type." So now we know.

There's a lot of competition in the magazine business.
And nowhere is this more true than among journals of personality. I mentioned that the candy study wasn't reported
in the *Journal of Personality*. It appeared in the *Journal of
Personality and Social Psychology*. I'll call these *JP* and
JPSP. There are others, too, like *JSP*, the *Journal of Social
Psychology,* and *JPA*, the *Journal of Personality Assess-
ment,* but you have to draw the line somewhere and I think
two of these journals are enough for anyone. *JP* is my favorite. I would put its fidgeting article up against anything
in *JPSP*, including the milkshake/candy article, and even
including the astonishing "Presidential Personality" paper
by Dean Keith Simonton of the University of California,
Davis, in the July 1986 issue.

To gauge the personalities of the thirty-nine U.S. presidents, Simonton, who calls his field "historiometry," gathered together biographical material—in other words, things
that other people had written about the presidents—and
put it all on index cards. Then, for each president, "raters"
read the cards (without knowing which president was being
described in a given set of cards) and decided which of 300
adjectives applied. Through methods too mysterious to re-
count here, the 300 adjectives were divided into fourteen
categories, each rated numerically to tell us things like who
was the tidiest president (Buchanan) and how many of the
thirty-nine showed less intellectual brilliance than Ronald
Reagan. (The answer, which says something about American history, or "historiometry," is twenty-five.)

Now, I know Simonton couldn't interview the thirty-nine

presidents, or preload them with milkshakes. Most of them are dead. And it would have been silly to have studied what they wrote themselves since these days a president would no more write his own speeches than he would iron his own shirts. I was, however, surprised to read that dozens of the "descriptors" from the list were *"prima facie"* useless because they didn't apply that well to any president. Among these adjectives were: cruel, fickle, foolish, obnoxious, and whiny. This made me wonder whether we were talking about the same presidents.

Whoever these guys are, Simonton's method produced some remarkable conclusions about them. Lyndon Johnson, who showed his abdominal scars to the world and had his advisers attend him as he sat on the toilet, was much tidier than John Kennedy. Eisenhower was enormously dumber than Ronald Reagan, who is just as smart as Richard Nixon. And Kennedy ties with Millard Fillmore for second on the most attractive. Franklin Pierce beat them both by a nose.

This is good stuff. And it may seem that between *JP* and *JPSP* it's hard to pick a winner, but that's only until you look at the advertisements. The best one I found in *JPSP* had to do with how to get grants. But on the back cover of *JP*'s love issue there was a pitch for psychological testing kits that brought back to me those great ads in *Boys' Life* and *Field & Stream,* the ones for taxidermy courses, body building, and worm farms. The *JP* advertisement was from Psychological Test Specialists of Missoula, Montana, which sounds to me like the place the taxidermy courses used to come from. The tests that looked good to me were "Proverbs Test (PT)," billed as a "Highly sensitive indicator of psychotic processes. $13.00 for kit" and the "Famous Sayings (FS)" test, also $13.00, in which "Agreement with proverbs, aphorisms, and folk sayings is analyzed to determine personality structure. Particularly useful in personnel selection." (All work and no play makes Jack a dull boy. Do you agree?) I'm very tempted to send away for them. The writing business is always shaky and I'd like to have a little something on the side. A little psychological testing, a worm farm, maybe some taxidermy—I think I could make a go of it.

Ecology Quinella

Remember ecology? Back when Earth Day and tie-dyed shirts and noncompetitive sports (was it hug ball they all played?) were in vogue, ecology was big. Of course, it's still big to ecologists, who never seem to tire of doing complicated mathematical analyses of the rises and falls of various aphids and beetles. (Remember Japanese beetles?) But as a buzzword, it has lost its buzz. It has been squeezed out of the public consciousness by investment banking and tax-free municipal bonds.

I suspect that the reason is the granola factor. I was talking to my seventeen-year-old nephew the other day and I heard him call some poor girl he'd met on a bus trip a "granola." I asked him what he meant. As you might expect, granolas are people who are overly devoted (in the opinion of non-granolas) to things like natural foods and organic gardening. By the process of metonymy, granolas have come not to be, but to be called, what they eat. I think that ecology, like my nephew's bus companion, smacks of granola, and granola isn't the thing to smack of in the eighties. Among today's celebrities there are very few granolas. Jello Biafra, who sings with a group called the Dead Kennedys, is almost certainly not a granola. Ronald Reagan and Donald Trump aren't granolas. Don Johnson doesn't even eat breakfast. I'm not a granola (of course I'm not a celebrity either), although I do have certain granola-like attitudes. I eat oatmeal, for instance, and I've been thinking about gardening organically.

Ecology may also be a "retro" science, but this I'm not sure about. I'm also not sure whether being retro is good or bad. I do know that it's a word worth looking into. The same week I heard my nephew use it, I saw it in the *New York Review of Books,* although there it wasn't in English. Someone was describing nostalgia as *le goût rétro.*

Certainly ecology, in its obsessive preoccupation with untouched nature, seems to have a taste for the past, and I don't mean the twenties and thirties, I mean the Mesozoic, when mammals were in no position to build power plants, and if there was any toxic waste it was presumably natural and in its proper place. The example of an ecosystem given to students is hardly ever the Bronx or Atlantic City. It's always that primeval pond. Remember the pond? Every student of ecology has had to hear about the algae and phytoplankton and fish and turtles. This pond is very old by now, also mucky, what my nephew would call rude. When human beings do enter the picture they always seem to be wearing a loincloth or toting wooden hoes. They live in villages, or travel in nomadic groups that treat the deer and the bear with the kind of totemic respect we reserve for the deutschmark and the yen. Perhaps the most famous of these villages is a hypothetical one that appeared in a parable by Garrett Hardin. He made it up to illustrate the problems of population growth and limited resources. In his pastorale, villagers share a common on which they graze their cattle. If they don't limit the growth of the herds all the cattle die. This is a good story, but out of date. Some cities, like Boston, still have commons, but grazing isn't what gets done on them.

If ecology is to regain its former position in society, it

needs a new image, something with a little snap to it, like mens' underwear for women. The image I have in mind is that of a racetrack. I think the racetrack could be the new controlling model in ecology. To explain why, I have a story to tell about racetracks, a kind of parable. Hardin called his the Tragedy of the Commons. I call mine the Comedy of the Horse Manure. Like most comedies, it isn't that funny to the people who are in it, but at least in this one the animals don't die.

Once, not too long ago, mushrooms thrived in the land of Pennsylvania, near Kennett Square. They were grown in the traditional medium of choice for mushrooms—composted horse manure. Fortunately there were a lot of horses in neighboring lands, at racetracks like Aqueduct and Belmont where the ponies raced and sharpies in white shoes placed bets at the $50 windows. As a natural consequence of all these horses the racetracks had enormous amounts of horse manure. People came up from Pennsylvania, manure haulers by trade, and paid good money for this manure, which made the racetrack people chuckle with delight. Of course the haulers chuckled too, because they turned around and sold the manure to the mushroom growers, again for good money. The growers, as you have probably guessed, also chuckled, because they sold their mushrooms for even better money.

One day, mushrooms from afar (Taiwan, China, Korea) arrived in cans and took over big chunks of the market for mushrooms in cans—the precise market in which many growers in Pennsylvania sold their mushrooms. Suddenly the money was not as good. The growers suffered, the haulers of manure suffered, and the tracks had to pay to have the horse manure taken away. Everybody at all ends of the manure business was unhappy, except one clever man who found a way out of this slough of manure and despond—a new way to sell the product. He turned it into compost right at the racetrack and sold it to garden centers, nurserymen, and landscapers so that it would grow not mushrooms but tomatoes, zucchini, privet hedges, and dwarf conifers. Once again the sharpies could place their bets at the $50 windows without undue worry about where the manure from all the longshots was going.

My parable happens to be true. The mushrooms are real. The manure is real. The track is real. It's the Saratoga Raceway in Saratoga, New York. This isn't the flat track where thoroughbred horses race every August and pedigreed humans show up in fancy straw hats and boaters. This is a harness track, open year-round, where the hat of choice is a baseball cap, preferably emblazoned with the name of some maker of heavy machinery. The clever man is Robert W. Morris. He didn't actually invent composting, but he did study it and experiment with different methods. He even went to visit what may be the world center of composting knowledge, the U.S. Department of Agriculture sludge composting operation in Beltsville, Maryland. Then, finally, he spent $225,000 to put together his own system. It turns the manure output of 1,100 trotters and pacers, which, mixed with the straw and sawdust from the stalls, amounts to 150 cubic yards a day, into rich, crumbly compost, what James Crockett, the garden guru, calls brown gold.

Before he set the system up, the track was paying $100,000 a year to get the stuff hauled away. Now it takes in $300,000 a year. Morris even mixes the clay and sand he scrapes off the track during maintenance with the compost to make topsoil, which he also sells. And he keeps down the flies by buying wasps that eat fly eggs—organic all around. Other tracks are following his lead. According to Morris, Keeneland in Kentucky has built a plant, Charles Town in West Virginia is building one, and a number of other tracks, including the Meadowlands in New Jersey, are studying the operation. Morris sells his plans and expertise for $10,000.

It seems to me that my parable, like Hardin's, has an ecological moral, which is that if we ran the planet the way Morris runs his raceway we'd be in a lot better shape. And we could do it. This isn't some imaginary village we're talking about; there are no aborigines living in an unspoiled wilderness in this story. This is spoiled wilderness—a harness track, not far from a major highway. This is no fantasy, it's a business. Here's an efficient, elegant, ecologically sound operation that was developed not out of a sense of oneness with all life, but for the sake of *money*. This is the heart of the harness track story in terms of public appeal. Nothing, but nothing, is less like granola than money.

I suppose there might be some question about whether the racetrack can replace the old pond as a model ecosystem. I don't see why not. A racetrack operation is built on the $2 bettor, whom we might compare to photosynthetic phytoplankton, or algae. In the pond the plankton and algae capture the energy of the sun and form the base of the food chain. At the track the $2 bettors do something similar. They provide the money. The parimutuel betting system, through which this money flows, guarantees that all life forms at the track are interdependent, as in natural ecosystems. If Coup de Fusil is 2 to 1 on the morning line but the bettors are putting most of their money on other horses, the odds increase to 4 to 1. Every bettor and horse is connected to every other bettor and horse.

In the pond, sunlight is transformed into vegetation, vegetation into animals, and those animals into bigger animals as one thing eats another. At the track, money is transformed into fancy cars, stocks and bonds, and more horses. Of course these stocks and bonds don't go to the phytoplankton. It's the large mammals at the top of the chain, the owners and (sometimes) the big bettors, who get rich. However, as with the familiar carbon and nitrogen cycles, something trickles back to the small fry to keep the system going. The $2 bettor may be largely excluded from the big money, but, at least at Saratoga Raceway, he can still get in on the manure. One hundred twenty pounds, already composted, costs just $5 at a local garden center.

Good scientific models have wide applications, and I think the racetrack could subsume the ecology and evolution of all life on earth. Science fiction writers are fond of speculating that the planet is a zoo maintained by higher intelligences. It works better as a racetrack. I figure the creatures who run it are strange energy conglomerations with cigars and white shoes who lounge around saying stuff like "I'll take the mammals in the Pleistocene," or "I like the hadrosaurs in the late Cretaceous." If this is true the question of interest to us has to be what the morning line is on human beings. I suspect we're a long shot, good in the first few furlongs but easy prey to a dark horse (or insect) like the cockroach because of our tendency to blow ourselves up in the stretch.

Still, long shots do come in now and then. Love Champ did. I couldn't visit the composting operation without going to the track, so on the day I went to Saratoga I spent the afternoon with the runners and the evening with the trotters, doing research. (I won two daily doubles and two quinellas and came out $40 ahead for the day. If you care to send me your money I'll be happy to invest it for you.) At the harness track, after looking over the past performances, I put $2 to win on Love Champ in the first. She surged ahead at the wire and paid 10–1. I also bought some Saratoga Organic that day, and later I worked it into my garden. The way I see it, Love Champ not only won me $20 but next spring she's going to make my tomatoes grow. Now that's ecology.

Planet of the Evangelists

ometimes I think the scientific establishment is too hard on creationists. They're not so bad. For one thing, they've got a sense of humor. The only way that I can see that the Bible can be literally true, and the planet thousands and not billions of years old, is that the entire geological column and the fossil record are part of a practical joke perpetrated by You Know Who. As jokes go, I like it. Imagine: a God who antiques planets just to confuse the inhabitants.

In literary terms we might think of this God as an unreliable narrator. And it's clear to me, if not to the creationists, that if He'd fool around with thousands of feet of rock just to trick us, He might also have stuck a few fibs in the Bible. Could it be that He's just teasing us about the seven days, the way He did with that bit about Jonah and the whale?

Looked at in this light the creation/evolution conflict is good fun—as conflicts go—certainly better than, say, a *thermonuclear* conflict, or the Super Bowl. This is the way I try to look at it anyway. I enjoy watching the creationists attack the fundamental basis of biological science and the evolutionists argue that whoever wrote the Bible wasn't versed in molecular biology. It tickles me to hear the claims on both sides about threats to morality, religion, the Constitution, and the minds and/or souls of our young people. Nonetheless, every once in a while, I get angry too. What gets me is when they start picking on the apes.

When *The Origin of Species* first hit the bookstores ape jokes were very big. Ape cartoons of Darwin were all the vogue, and the pugnacious T. H. Huxley, who often served as Darwin's front man, was asked, at a meeting of the British Association in 1860, on which side he was descended from an ape, his mother's or his father's. You'd think that things would have changed, now that apes have learned language, and star in one television program after another. But they still get no respect. Just recently a witness in one of these endless creationist court cases (this one is on its way to the Supreme Court) came up with this tired old saw: "I think if you teach children that they are evolved from apes, they will start acting like apes."

Well, really. Never mind that apes and human beings are actually descended from another creature. That's just evolutionist nitpicking. The point is that we're obviously related to apes, as anyone knows who has ever taken a good look at a chimpanzee's fingers. It's equally obvious that we act like them and always have, no matter who or what we think our ancestors are.* It's easier now than ever before to prove this contention, because we have definitive information on precisely how at least some apes act. We have the recently published and thoroughly impressive *The Chimpanzees of Gombe* by Jane Goodall, a 673-page masterwork in which Goodall does for Gombe what Joyce did for Dublin.

The chimps have their bad moments, it's true—murder, rape, infanticide, cannibalism—but nobody's perfect. On

---

*And sometimes we act *with* them, as in *Bedtime for Bonzo.*

the positive side, they usually care for their children, are very social, engage in coordinated group hunts, and have brief sexual affairs in which the male and female go off together for a weekend or more to the Gombe equivalent of a country inn—a tree perhaps, or a glade. The chimps form friendships. Brothers and sisters, and brothers and brothers, forge lifelong bonds. They're always grooming each other, to the point they sometimes seem to be an entire species of hairdressers. And most impressive of all, they have a rudimentary political system. The group is dominated by an alpha male, who's periodically challenged by aspiring leaders. Traditionally, leadership is gained and maintained by strongmen (strongchimps?) who form coalitions with other apes. There are also times of instability, when no ape reigns. It's just like South America.

If this isn't enough to prove how much alike we are, consider the astonishing evidence among the Gombe chimps of rudimentary shopping behavior. For instance, they test fruits they're about to eat by squeezing and sniffing them. And once they've brought a piece of fruit home, or at least picked it, they have what you might call recipes, established ways to prepare and eat different kinds of food. They don't write these recipes down, but we can.

Fig Leaves

Pick the leaves one by one. Pile together. Fold them over and chew well.

Aspilia Leaves

Press and rub against the palate. Do not chew.

These may not be up to Escoffier's standards, but if you've ever seen a dog eat you know that by comparison chimps are well on the road to the perfect omelette. They even have cultural differences in cuisine. The Gombe chimps feed on various parts of the oil-nut palm. But a nearby group, the Mahale chimpanzees, won't touch it. Goodall doesn't suggest that this is the result of a religious prohibition, however, merely a matter of taste and custom.

I know what the creationists are going to say to all this. They're going to say that I've forgotten something of supreme importance. Human beings have souls and chimpanzees don't. Well, a soul is a hard thing to measure. On a more concrete but still soul-related level, human beings also have churches and preachers. I can't honestly say that chimpanzees go to church (in the wild), but preachers are another matter. I can show that preachers and chimpanzees are quite a lot alike. In fact, since Goodall has given us abundant evidence in her opus of the individuality of each ape, I can go beyond this generalization. I can say which preacher is like which ape.

Remember now, I'm not saying these guys are chimps, just that they're like chimps in certain ways, almost all of them complimentary. It so happens the preachers I have in mind are both well-known television evangelists. This is only fair. If you're going to compare religious leaders to chimpanzees you've got to go for the big ones. I had thought of doing the pope, but Roman Catholic theology is so hospitable to evolution that it takes the fun out of it. As far as we know, John Paul II would have no objection to being linked (biologically) to other primates.*

---

*Darwin himself was almost a preacher. During his youth he sowed his wild oats by studying for the ministry. Of course he later settled down and became an evolutionist, or perhaps I should say *the* evolutionist. Darwin's father (the ministry was his idea) must have seen it coming. He said to his son once, "You care for nothing but shooting, dogs, and rat-catching, and you will be a disgrace to yourself and all your family."

I'm not at all sure that the two preachers I've picked will take the comparison with the same good grace one expects from the pontiff. They're Jimmy Swaggart and Pat Robertson. The latter has the added attraction of wanting to be President, which is very similar to being alpha chimp in Gombe. Jimmy Swaggart recently gave his support to Robertson's presidential quest, but that's not why I picked him. I picked Swaggart because he's my all-time favorite television evangelist. I watch him whenever I get the chance. Jimmy Swaggart is running for prophet, not President. He sweats, shouts, calls out the evolutionists, and has a stage presence that would've done Elvis proud. Now, we all know that chimpanzees jump around, thump their chests, and run at each other. Of course the chimps do this to show how tough they are, while Jimmy Swaggart swaggers and shouts to show how tough God is. Nonetheless, if you read Goodall you'll be compelled to view Jimmy Swaggart as the human equivalent of Goliath.

To quote Goodall. "He [Goliath] was aggressive and had a very fast, spectacular charging display, coupled with an unusually bold disposition. If the adult males were confronted by some strange object (such as a dead python), Goliath was generally the first or only male to approach closely." You shake a dead python at Pat Robertson, he might hesitate. Jimmy Swaggart is going to jump up, praise the Lord, and rip that snake right out of your hand. Then he's going to let you have a few choice verses about serpents—in the teeth. Goliath was in his time an alpha male, so I hope Jimmy Swaggart will see the comparison as a kind of honor, and not take offense at being compared to an ape who was named for an unbeliever.

Robertson is a bit more difficult because of his lack of emotion. I'm not saying he's completely without life. He's just—well, he's the Merv Griffin of religion. He smiles no matter what he's saying. He's perfectly suited to television, and television has brought him to prominence. He developed the Christian Broadcasting Network, and is the host of its flagship show, "The 700 Club," which is a lot like Merv except that David Brenner is never on. It's Robertson's mastery of the cool medium of television that suggests which chimp he's like: Mike.

I didn't pick this chimp just to work an old Pat-and-Mike (or Pat-is-Mike) joke. Among the chimpanzees of Gombe, Mike did much the same thing Robertson is attempting to do. He was cool. He never actually attacked anybody. He got to be the alpha male by mastering a new medium. Of course in his case political prominence came from the mastery not of the airwaves but of empty kerosene cans. Mike could charge down a hill, tumbling these awful-sounding cans before him, and scare the pants, if they had any, off every ape in the neighborhood.

I'm not suggesting that just because Mike got to be president of the Gombe chimps, Robertson is going to get to be our President. In some ways, people are considerably more clever than chimpanzees. All I'm saying is that Pat Robertson is a chimpanzee's relative, and acts like one, as do we all. (I want to point out that I didn't use the word uncle, and that not even the creationists confuse monkeys and apes.)

There's a political as well as a scientific lesson in all of this. It's always important, whether you're a human being or a chimpanzee, to look beyond the kerosene cans and the camera angles. For instance, I myself am a single-issue voter. My issue is apes. The apes are my relatives, I'm proud to have them, and I don't intend to vote for anybody who isn't related to them, or is ashamed to admit it. I have one question for each presidential candidate, and I think it's a question everybody should want to know the answer to: Are you kin to the apes or not? Once we know, we'll know how to vote.

This Little Piggy

They wouldn't let me see the micro pig. It may seem that this is of importance only to me, and perhaps to the pig, but I think we all have a stake (or perhaps a chop) in this conflict. The public has a right to know about the micro pig. As a member of the press, if not the working press, I represent the public. If I don't get to see the pig, who does?

It's not as if "micro pig" were the name of some new missile. We've long since passed the stage when we gave cute names like Fat Man to nuclear weapons. (If there were a Nobel Prize for bad taste, surely the guy who came up with that one would win it.) These days technology is dominant, and it provides, rather than suffers, nicknames. Consider the Refrigerator, the first home appliance to play in the Super Bowl.

No, the micro pig isn't a bomb. It's a pig. A tiny pig. Of course you can't call it a tiny pig. Then it would sound like something from a children's story instead of a Swine Research Laboratory. In science, there are no little piggies. When pigs get little, they become mini. When they get really little, they become micro. Soon, no doubt, this gradual swine reduction program will culminate in the long-awaited laptop pig.

In actual avoirdupois, the measuring system tradition-ally favored for pork, an adult micro pig weighs 50 to 100 pounds. A mini pig weighs 150 to 200 pounds, and a serious, old-fashioned, mainframe pig goes 600 to 800 pounds. My own personal pig equivalence chart produces the following equations: One regular pig equals two Refrigerator Perrys and change. One Yucatán mini pig (that's where they come from) equals one place kicker. One Refrigerator equals at least three micro pigs. Of course these comparisons aren't really fair. In the NFL, pigs don't get to carry the ball; they are the ball.

Still, tinyness—or micro-ness, if we must—is no reason to keep the pig hidden. What could there possibly be about a small pig that would make someone refuse to show it to me? This isn't Mikhail Gorbachev's tiny pig. It's not Deng Xiaoping's pig, destined for tiny, Communist moo shu pork. It's not even the Pentagon's pig, unfortunately.

Can you imagine what someone could charge the Defense Department for a micro pig? It could change the economy of rural America. But that's wishful thinking. This isn't a war pig. This is a peace pig, bred for medical research. One of the rules of science is that the smaller the experimental animal, the better. Hence the popularity of fruit flies and mice. At 800 pounds, the classic pig is unwieldy, to say the least. The Yucatán mini pig, which nature itself produced, was a step in the right direction. Then came the micro pig, developed through selective breeding by Linda Panepinto at the Miniature Swine Research Laboratory of Colorado State University in Fort Collins.

As soon as I read about this pig, I wanted to see it. I called Panepinto to set up an appointment. I was frank with her. I told her that I wasn't the least bit interested in the pig's usefulness to research, in how it was bred, or in whom it would cure, but only in how small it was. Panepinto was not overly enthusiastic about this approach to her work, but she was game. I made my airplane reservations and checked the map to see where I could go trout fishing when I was out there. (How long can you spend looking at a pig?) I sharpened my pencils and packed my fly rod. And then she canceled. I was aghast. The powers-that-be in the university, not Panepinto herself, had ruled that I was not to see the pig. They didn't want publicity—at least not from me. The reason? According to Panepinto, they were worried about animal rights activists. I suppose they were afraid I'd write some headline like: "Incredibly Cute Tiny Pig Destined for Vivisection."

Of course I never planned to write a headline like that. The thought had never crossed my mind. At least not until that telephone conversation, not until I felt the meat locker chill and heard, in Panepinto's voice, the dull thud of the Pork Curtain as it fell, encircling Colorado State like some vast and impenetrable side of ribs.

This has to give anyone in the press, or the pig business, pause. It raises certain questions, first among them being: How big is the pig really? I've seen a picture of one posed with a young woman, a researcher at Colorado State, but there's such a thing as trick photography. And even if the photography is accurate, who's to say that we're looking at a small pig and not at a giant person? That would explain the university's nervousness. There are restrictions about experimenting on people. You can't grow a scientist who's twenty feet tall and expect the National Science Foundation to look the other way. But I doubt Colorado State would try that. Even behind the Pork Curtain you couldn't keep a twenty-foot-tall woman a secret.

So let's stipulate the tiny pig, and assume widespread fear at Colorado State that the animal will be perceived by certain groups as just too cute for research. That's still not a reason to keep me away; I'm not part of the animal rights movement. It would be disingenuous of me, having made my peace with roast pork and barbecued ribs, to complain about a few experiments (as long as they use mesquite wood). And besides, as I understand it, these pigs (there are now a number of them) are useful because, among other things, they're susceptible to ulcers and atherosclerosis. Human beings stand in line at restaurants waiting to pay to get atherosclerosis. This is hardly what I'd call vivisection.

No, what I think Colorado State was afraid of was that I would make people understand how small a micro pig really is, and how contrary petiteness is to the nature of pigs. Somebody at Colorado State must have realized that if I were allowed to see the micro pig, and tell the world about it, I would have been able to set off a groundswell of protest and outrage that would have swamped the Miniature Swine Research Laboratory. You see, my beef with science (I've got a regular meat loaf going here) isn't that they use animals in experiments, but that they make them look silly. My position is this: Experiment on them if you have to, but leave them their self-respect. If I ever get anyone else to join me in this position, I plan to call us the Animal Dignity Movement.

It's not just pigs. Think of all the poor fruit flies with crinkly wings. And mice. I don't mean the millions we kill each year in research. I kill almost as many in mousetraps in one cabin in upstate New York. (I figure dying in traps is a noble end for mice. And noble or not, if they want to leave turds in my silverware drawer, they'll have to take their chances.) The fates worse than mousetraps that I'm thinking of involve not pain but ugliness and incompetence. Scientists have bred super-obese mice that can hardly move, naked mice that have no fur, mice that can't smell, mice that can't walk a straight line. You can even cross some of these strains and get, for example, a fat naked mouse. I hate those.

You can't build a mass movement around the humiliation of flies and rodents, but pigs are different. People care about pigs. There's a whole pig *thing* in this country. There are piggy banks, and porcelain pigs that aren't banks, not to mention Miss Piggy and the swine flu. People have been called pigs for their table manners and their politics, sexual and otherwise. In 1968, in Chicago, a swine named Pigasus was nominated for President. Some people keep pigs as pets, and other people race them. Racing pigs are set loose on a track to gallop, not for the gold, but for an Oreo cookie. Pigs today are probably the most highly cathected of farm animals.

And what, of all piggy characteristics, has given the animal such a secure seat on the *zeitgeist?* The answer is size. I've seen pigs in the pork—real, mud-covered, adult pigs— and they're nothing if not huge. They're immense. They're fat. They're round. They snort. They are, in a word that incarnates the culmination of swinehood to which every piglet aspires, hogs. That's the true destiny of pigs, but it's a destiny denied to the little piggies at Colorado State. At fifty or so pounds they might make the cutest little Canadian bacon anyone has ever seen, and help cure heart disease to boot, but they can never, ever be hogs. I find that sad.

If I'd been to see the micro pig, I might have been able to make something out of all this. I could have really stirred things up. Alas, it wasn't to be. The people at Colorado State saw me coming. They put a lid on me, all right. It's not fair. Colorado State is a public institution that uses a lot of public money, and it shouldn't shut science writers out of the pig process. But I don't know what to do about it. Maybe a letter campaign would help. Maybe it would be good if there were a spontaneous outpouring of support (for both me and the micro pig) in the form of telegrams and letters.* After all, it's not only the First Amendment to the Constitution that's at issue here. They ruined my fishing trip.

---

*The address is: Department of Public Information, Colorado State University, Fort Collins, CO 80523.

Paradise Lost

I had a few antiques shipped to me the other day. One of them, a Calville Blanc, was a very popular item in 1620, during the reign of Louis XIII. Another was early American, of a type favored by Thomas Jefferson. Those in the trade call it an Esopus Spitzenberg. This is only the beginning of my acquisitions. I plan to get more soon. For example, I have my eye on a nice Edward VII, and when I get it, I'm going to do just what I did with the first two. I'm going to eat it.

It's O.K. They're apples. Louis XIII didn't sit on his Calville Blancs, he had them for dessert. The Esopus Spitzenberg, a fruit of agrarian democracy, was Thomas Jefferson's favorite apple. Every time I bite into one I imagine myself writing the Declaration of Independence. I also pen a note to King George III to go along with his copy: "Dear King, How do you like them apples?"

The Edward VII is another old-fashioned apple, what the purveyors call an antique, although apples aren't antiques the way desks and chairs are. Individual apples are grown right now, in our own time—the present. It's the varieties that are old. And, like old furniture, they're disappearing. The same is true of many old varieties of vegetables, like the famed Jenny Lind muskmelon—as far as I know the

129

only vegetable ever to have been eulogized in the pages of *The Wall Street Journal*. It must have been good.

I ordered the apples so that I don't have to read their obituaries, too. Once I decide which ones I like, I'm going to order trees and grow my own, right next to the garden with the heirloom vegetables (that's what the old-time varieties like the Jenny Lind are called). For the spring I'm thinking of Low's Champion shell beans, Rocky Ford honeydews, Pike rutabagas, and that good old-fashioned, non-hybrid, open-pollinated Golden Bantam sweet corn.

I'm not alone. This is a trend. I know because everything I do is part of a trend. I was born in 1949, which was, in itself, trendy. In the sixties I had long hair and went on peace marches. Later, while still young, I moved to an urban area and became a professional (sort of). It's demographic destiny; if I'm starting to buy old-time apples and heirloom rutabaga seeds, the whole culture is on the edge of, or perhaps already in, the era of the biological antique. The passion for old objects is about to reach, in my generation, its ultimate form. Forget chairs and tables, porcelain and silver. We've combined our taste for old furniture with our love for herbal tea and hiking clothes, and we're going after the *natural* past. We want the old genetic furniture from the planet's attic (the good pieces, of course). "Give us that old-time tomato" is our hymn, and you can bet there's somebody ready to sell it to us, with a kilo of amaranth on the side.

Even the government is involved. It has a National Seed Storage Laboratory, where more than 200,000 varieties of seed are preserved in a kind of giant Burpee's Memorial Garden Museum. (The lab, coincidentally, is located in Fort Collins, Colorado, the home of the famous, but elusive, micro pig.) Home gardeners do something similar through the Seed Savers Exchange, growing and preserving seeds from disappearing cultivars. And there are commercial operations as well. As with all the other things my generation loves, antique fruits and vegetables have inspired mail-order catalogues. These fairly throb with historical romance. My favorite is from Southmeadow Fruit Gardens of Lakeside, Michigan, which carries the Lady Apple, "known in Europe as Api or Pomme d'Api . . . Some writers even trace its origin to Appius Claudius, the Roman censor who constructed the Appian Way and who is asserted to have brought the apple from Peloponnesus." From the Peloponnesus—can you imagine?

My vegetable seed catalogue, from Johnny's Selected Seeds, of Albion, Maine, is more restrained. But the name of the town does suggest, if not Arcadia, at least pre-Thatcher Britain, and the text hints at an unspoiled American past. Rhode Island White Cap, for instance, is described as "an authentic, 8-rowed white flint corn of the Narragansett Indians." This is a good thing to have around in the era of monoculture. Because so many farmers plant the same varieties now, their crops are more susceptible to disease. The 1970 corn blight wiped out a fifth of the nation's corn crop, but I'll bet it didn't hit Rhode Island White Cap. Presumably, by growing it, I can kill two birds with one stone ground cornmeal muffin. I preserve genetic diversity, and I get to pretend I'm a Narragansett Indian in pre-Columbian America. Of course it doesn't pay to think too hard about what happened to the genome of the Narragansetts themselves.

Such are the pleasures of biological nostalgia, pleasures that I should point out are distinct from those of the scientific discipline closest to antique collecting—paleontology. It might seem that paleontologists go after the real biological antiques, and it's true that their subject is the history of life. But they only collect remnants. Usually you can't eat what they find. And the nature of their lust for the past is fundamentally different from the urges that send a person in search of Ashmead's Kernel (a great old russet apple that I got from Southmeadow Fruit Gardens). Paleontologists do not want to live in the past. They just want to learn about it, and what they've learned is that it wasn't so great. There was one mass extinction after another, comets kept falling, there was always some new predator evolving, volcanoes were a constant danger, and there were these godawful spasms of orogeny. (I think talk like this is why geology has become so popular recently. It's sex without guilt—in fact, without sex.) The rest of us aren't so clearheaded. We seek out things from the past because we don't like a present in which supermarket tomatoes are designed to survive global thermonuclear war and we're not.

Nowhere is the underlying nature of biological nostalgia clearer than in the popular fascination with the ice ages. This doesn't have to do with collecting antique animals, exactly; that's very hard to do. It has to do with antique varieties of people. Ice Age Person is very popular right now. In addition to movies and books (*Iceman* and *The Clan of the Cave Bear*), there was a recent exhibit of ice age art at the American Museum of Natural History in New York called, romantically, "Dark Caves, Bright Visions." It had lots of great jewelry, art that I could understand, and I learned from it that these people not only had furs, they had designers. As the book based on the exhibit puts it, "Several of the 110 or so engravings of humans from the 15,000-year-old Magdalenian site of La Marche in France . . . seem fully dressed in tailored clothing with cuffs and collars." The fall line for 13,000 B.C.

I think the exhibit offered reason enough to yearn for Magdalenian times. There were lots of big mammals to spear, low population density, Sundays in the cave with Giorg. But somebody always has to come up with a new, improved version. *Newsweek* did a cover article on the ice ages. It ran a picture of Ice Age Man. He was Don Johnson.

He didn't have his T-shirt on, but I recognized him anyway. He had the right amount of beard, and he was displaying his trademark dullard's pout. There was even a young woman in lynx, or maybe fox, holding on to his arm and leaning her head on his shoulder in what looked (to the layman) suspiciously like a 1950s romantic stupor.

I would consider this an isolated incident of wishful paleoanthropology except that the woman looked as if she might be related to Daryl Hannah, who starred in *The Clan of the Cave Bear*. You may recall her as the debutante raised by Neanderthals. As an actress, Hannah seems to specialize in the extreme boundaries of the human species; in other movies she has played the roles of mermaid and performance artist. I like Daryl Hannah, even if the Neanderthals didn't. (They thought she was incredibly ugly, which is probably why they went extinct.) But like her or not, I have to say she didn't belong in *The Clan of the Cave Bear*. It makes the death curse of the Mog-ur so much less scary when the foremost question in your mind is not whether Ayla will survive, but how she kept her skin so soft and white.

I don't doubt that life in the ice ages was great, probably better than in the 1950s, which I remember as one long effort not to get beaten up. But I'll bet my totem (House Mouse) that there was nobody in the paleolithic with Daryl Hannah's complexion. I know who was there. I've seen those sculptures of Ice Age Woman. You know the ones I mean— the *big* ones. I know all about the eye of the beholder, but I think few modern beholders would be prepared to date the Venus of Willendorf. As for the men, they were human, but they didn't look like Don Johnson; they looked like George Shultz.

I say we stick to fruits and vegetables. They really did taste better in the old days. And if we want to pursue a real golden age, there's no better way to do it than with apples.

We could go back before the ice ages; we could go *all* the way back, to the time when nobody had to work, nobody had any clothes, and there was no original sin. Why not? Why not go after the ultimate biological antique, the original apple? That's what I really want growing in my yard. I can taste it already, sweeter than a Delicious, prettier than a Pomme d'Api, crisper than a Calville Blanc, big, russet, richly flavored, and forbidden, the favorite apple not of Thomas Jefferson but of God.

Science "Я" Us

I now have in my house two microscopes, a chemistry set, a geology lab, a gyroscope, Uncle Milton's Giant Ant Farm, and a bunch of dead Sea Monkeys in a Micro-Vue Ocean Zoo. I acquired this technical and zoological menagerie because of a flaw in my character. I'm an evangelical consumer. This doesn't mean I buy a lot of Bibles. It means that I believe in salvation through shopping. Like most religious beliefs, this one isn't rational. In actuality, I'm continually disappointed. Nothing, not even the new Walkman or the pre-abused blue jeans, has brought quite the happiness I hoped for. But then, not everybody gets cured at Lourdes either.

My last shopping episode, one might almost call it a seizure, occurred just before last Christmas. I was in Toys "Я" Us buying gifts in my usual way—one for them, one for me. I turned down the science toys aisle and there before me stood shelves brimming with faux chrome microscopes, plastic models of the human skull (always big at Yuletide), chemistry sets that promised hundreds and thousands of experiments. "Scientific Fun," the print on the packages blared, and "Research Set," and—my favorite—"Here AT LAST is the Miracle of Nature That Goes BEYOND The WILDEST DREAMS OF SCIENCE." Suddenly the fluorescent light seemed to acquire a revelatory luminescence, the kind of blinding glare that would have struck Paul if it had been the subway he took to Damascus. There it was, all boxed up for me, the joy and struggle of intellectual pursuit. I had heard that you couldn't buy happiness, but nobody had ever told me you couldn't buy science. I bought it, and I took it home.

Then I opened it up. The awful thing about gifts is that you never really get what you want, even if you buy them for yourself. With the first few scientific disciplines I suffered grave disappointments. Gyroscopy, for example, turned out to be kind of limited. Once you've balanced the thing on top of a pencil, you've pretty much plumbed the depths of excitement in that field. Geology wasn't much better. In fact, the geology research kit I bought, for $19.97, was considerably less interesting than the gyroscope. This isn't what John McPhee had led me to believe. John McPhee is a writer who treats geology as if it were something incredibly interesting, like Cajun food. He writes about basins and ranges, mountain building, plate tectonics, and volcanoes. He reported what I consider to be the only interesting thing ever said about golf, which is that golf courses are human attempts to reproduce the glaciated terrain of Scotland—where the game began.

There was nothing this interesting in my research set. What I had paid $20 for was a collection of twenty-four pieces of rock, most of them fairly mousy (slate, sandstone, lava, and of course the ever popular Piedmontite schist), a magnifying glass, and a seven-page booklet. My guess is that it wasn't written by McPhee. I quote, "No. 6 Gabbro (plutonic rock). Consisting mainly of pyroxene and feldspar rich in calcium, it is a black rock with rough grains." I guess this is what makes kids love science.

I turned from rocks to raisins. In my chemistry kit I

discovered an experiment that I call the "Lazarus Raisin." This is what you do: drop a raisin into a glass of seltzer or soda and tell the raisin to go up and down. I did this with my kids. I said, with feeling: "Rise up, raisin!" and "Go down, raisin!" as the raisin rose and fell, borne by seltzer bubbles. My two-year-old got into the spirit of this. She raised her arms and shouted like a faith healer when the raisin lay too long, apparently paralyzed, at the bottom of the glass of seltzer. "Rise up raisin! Rise up raisin! Rise up raisin!" she called out in her best basso profundo, stamping her feet, stretching her hands to heaven, or perhaps to the ceiling. Unfortunately, she always got bored before the raisin got cured, so she never saw the actual miracle take place.

Obviously, we had hit upon the perfect Bible Belt soda commercial. But something was missing. I wasn't experiencing what Einstein experienced when he thought up general relativity. At least I hope I wasn't. Shouting at the raisin didn't have the feeling I was looking for. It didn't feel like the search for the structure of DNA. It felt more like a new feature on the David Letterman show—stupid food tricks. I suppose I could have gone on. I could have made borax crystals. I could have put lemon juice in milk so that it curdled and produced casein. I could have used the casein to make paint. I could have painted something. My heart wasn't in it.

My heart, if truth be told, is and always has been in ethology. I've always longed to be out in the field, observing African elephants, or spending a year with the giant river otters of Brazil. The animal world is so rich, the discoveries so varied and intriguing. I still recall my astonishment at learning that when two related elephants meet each other after an absence of weeks or months, they sometimes charge together and enter into a kind of ecstatic embrace, which makes them both so excited they tend to urinate and defecate in the process of saying hello. When I first read about this it confirmed my belief that it's neither language nor intelligence but manners that set human beings apart from animals.

Unfortunately, I've never been able to get the funding to go to Africa or South America. I did, however, have the $6.97 I needed to buy the "Miracle of Nature That Goes BEYOND etc., etc." Beware the inexpensive miracle. In this case the miracle is, or are, sea monkeys, which also go under the less transcendental name of, yes, Virginia, brine shrimp. Of course these are bigger than average brine shrimp and they're supposed to live longer, too. It seemed reasonable to assume they might make an appropriate subject for a household ethologist. And perhaps someone whose animals lived to participate in the Super Sea Monkey Race or the forthcoming Sea Monkey baseball would have more to report on in the way of behavior. (Both of these activities take advantage of the classic brine shrimp tendency to swim upstream.) Mine lived their short lives without the pleasures of sport, as far as I could see. They hatched all right. And they took up swimming around their container, little dots with the pale color of blind cave catfish. They didn't look like pets to me. They looked more like intestinal parasites, the sort of things you might find wriggling around in a glass of the Nile River. Despite my best efforts, they lived only a week, and in all honesty I can't say I was sorry when they died.

I would never say that about *Pogonomyrmex californicus.* These are my ants—the ones in Uncle Milton's Giant Ant Farm. These are the creatures that finally provided some behavior for me to study. You can't go wrong with ants. Whether you've got them in a homemade formicarium or in Uncle Milton's farm, with the little silo and windmill, they're still insects of distinction. For example, ants are the source of the single most problematic word in the English language: "formication." It means "an abnormal sensation resembling that made by insects creeping in or on the skin." Dermatologists learning English as a second language face the gravest difficulties, but because of the obvious perils, this word is seldom taught even to those learning English as a first language. Instead we learn the colloquial "creeps," as in "he gives me the creeps."

So few creatures are wonders of etymology *and* entomology. Ants are tops. Or, as E. O. Wilson of Harvard, a major figure in formic science, says, "Ants are in every sense of the word the dominant social insects." Besides which, you can get them through the mail. I've been watching my ants now for a couple of weeks, and the thing I like best about them is that they're still alive. They're digging tunnels, and making hills around the silly little silo. They're eating and bustling about. They work; I watch. And I've made a couple of discoveries, one of which is that they don't know how to swim. I put too much water in the farm one day and it made a little puddle. I watched, stunned, as one of my *Pogonomyrmex* tried to walk over the water, stumbled, and fell in head first. I thought it was drowning. I was going to save it until I remembered it was an ant. I'm fairly sure that ants don't breathe through their mouths, and while I was thinking about exactly how they do breathe, this one managed to save itself.

Another notable thing about ants is that they don't panic when something awful happens to them. I say this not only because of the drowning false alarm (there was a bit of agitated feeler-waving in that incident) but because of something I witnessed in the graveyard detail. When the ants aren't digging or eating, they're carrying dead ants, or parts of dead ants, off to some nether corner of Uncle

Milton's farm.* I worry a lot about this, given the plague that struck my sea monkeys. But I've been impressed by their seriousness. Several ants were, shall we say, bruised, as I transferred them from the shipping vial to the farm. One of these ended up without the back half of its body. I saw this ant, insouciant, carrying a piece of a dead comrade, or perhaps of itself, off to the dump. This is the kind of animal that Vince Lombardi would have loved.

I'd like to end this piece on a happy note, with me in front of Uncle Milton's Giant Ant Farm pursuing my ethological studies, happy, for once, with my purchases. However, as the days passed an ethical crisis arose. Did I mention the microscopes? I bought one that looked like chrome and was supposed to go up to $1200\times$, with slides, tweezers, and prepared slides, for $42.97. When I tried it out, everything at every power was just a rainbow pattern, the kind of kaleidoscopic image people liked to look at in the '60s when they were smoking marijuana. It was a terrible disappointment. The pollen, the cat hair, the bumblebee leg—they all looked alike.

I ordered a better, more expensive microscope. This one only went up to $300\times$, and it didn't have any kit, but it cost a hundred dollars. It took a very long time to come. Finally, it arrived, on December 24. It was beautiful, made out of metal instead of plastic, with achromatic correction to keep the rainbows away, and its own wooden case. And it worked beautifully. I tried all the prepared slides that had come with the other kit. The sunflower pollen looked like little burrs, and the tulip pollen like chips of semi-

*His full name is Milton Levine and he incorporated, as Uncle Milton, Inc., in 1954. He started selling his ant farm on July 4, 1956, and has sold, up to now, about 8 million. Most of them are pretty good, but the one I had as a child never worked for reasons that have always puzzled me. Recent conversations with my parents revealed to me that we may, in fact, have never sent away for the ants. That would do it.

precious stones. The bumblebee leg with its pink stain was wonderfully clear, and hairy. What to look at next? Well, if a bumblebee leg was fascinating, a *Pogonomyrmex* leg would be too, wouldn't it? Or perhaps a whole *Pogonomyrmex*.

I was beset by conflicting voices. One asked me whether I was an ethologist, with respect for my subjects as individual creatures, or a reductionist—I mean this literally, in the sense of reducing a *Pogonomyrmex* to legs, thorax, and abdomen. Jane Goodall wouldn't dissect her chimps, after all. The other voice asked me if I was an idiot. It pointed out to me that we weren't talking about primates. We weren't talking about mammals. We were talking about a creature that didn't notice when half of its body was gone. It also pointed out to me that, according to E. O. Wilson, "at any given moment there are at least 10^{15} [in layman's terms, a zillion] living ants on the earth." It didn't know whether this included half-ants.

I don't know either. But I blame what happened on shopping. As long as I had had only the rotten microscope, the ants were safe. What was there to see? But I had to have a new, better, more expensive microscope. And as any scientist knows, once you've got the instrument, you've got to use it. (I consider what happened to me and the ants to be an excellent argument against nuclear weapons research.) The images were so clear with the new microscope. I couldn't help but imagine a *Pogonomyrmex* at $50 \times$—the polished amber abdomen, the intricate mouthparts, the delicately furred feelers, the faceted ebony eyes. I don't want to go into the whole sad story. Suffice it to say that the reality went BEYOND My WILDEST DREAMS, and there were $10^{15} - 1$ living ants on earth last Christmas day.

A Gastronomer
in Paris

*The discovery of a new dish does more for human happiness
than the discovery of a star.*

—Jean Anthelme Brillat-Savarin

T rue wisdom doesn't fade. That observation was made
in 1825 in Brillat-Savarin's brilliant *The Physiology
of Taste*, a book that I might point out was published
more than a quarter-century before Darwin's *Origin of
Species*. And it's as valid today as it was then. Not that
either I or Brillat-Savarin has anything against astronomy.
The point was merely to beef up another noble science,
gastronomy, and to give comfort to its practitioners, the
gastronomers.*

Brillat-Savarin's work would be, and in fact is, a classic
in any field. It's a book of meditations in which he treats,
with equal seriousness, the "Theory of Frying," "The Origin
of Sciences," "The Pullet of Bresse," and "The End of the

---

*You may find, in food magazines, that the term "gastronome" has
been lifted, undigested, from the French. If we had an Académie Amér-
icaine to defend our language and culture, I'm sure they would object.
After all, is Carl Sagan an astronome?

World." This is as it should be. Brillat-Savarin's only mistake was in thinking that the world would recognize, as he had, the supreme importance of good food. He foresaw a great academy of gastronomy, on the order of the Sorbonne, or Harvard, or Berlitz, which would make the name of its founder famous. "It [the name] will be repeated from century to century with those of Noah, of Bacchus, of Triptolemus . . ." Ah yes, Triptolemus.*

It should be obvious that Brillat-Savarin was French. Here, after all, is a man who devoted his life not only to eating, but also to *thinking* about eating. Only a Frenchman would enumerate the four, and only four, classes of people who eat, God forgive them, bouilli (beef that has been simmered to make stock, or bouillon). Only a Frenchman could say, "Fish, by which I indicate all species of it considered as a whole, is for a philosopher an endless source of meditation and astonishment."†

Of course that's what the French are for, isn't it? That's why we have them, to produce great food, and a certain sort of exotic genius. They, their cuisine, and even their insufferable self-importance are a kind of international resource. Or so I like to think. To be fair, I would have to admit that this image of the French is a stereotype like any other. And I would have to say also that this stereotype is being promulgated by a man who has spent more time in Madawaska, Maine, than he has in Paris. (This isn't as bad as it sounds; Madawaska, although it doesn't have the Louvre,

*The goddess Demeter taught Triptolemus the arts of agriculture, which he spread throughout Greece.
†In reading Brillat-Savarin, I relied on the translation of M. F. K. Fisher, whose notes seem to me every bit as good as the text. She says in her footnote to the fish comment, which I reproduce here in part (in my fish footnote), "The most meditative statement I have ever read about Fish, not a fish or the fish but Fish, is a poem from the Japanese." It is: "Young leaves ev'rywhere/ The mountain cuckoo singing;/ My first Bonito!"

is mostly French. Its people are the Acadians of Longfellow's *Evangeline,* who, transplanted to Louisiana, became Cajuns.)

You can see that I'm deeply involved with the French, although largely in my mind. Probably the best way to put my relationship with them is to say, as Woody Allen did about a blind date (he's lucky he wasn't involved with a Frenchwoman), "I *really* have mixed feelings about [them]." I know that not everyone shares these feelings. Some people just plain don't like the French. No doubt these people are all Class 3 Bouilli-Eaters: "The uninterested, who not having received the sacred fire from heaven, look on meals as a duty to be performed." But for those of us who have a taste for culinary art (and paintings of water lilies—between meals) the French are, well, they are to us what "Fish" is to a philosopher—an endless source of meditation and astonishment.

That said, it should be clear why it depressed me to receive in the mail my issues of FAST. The acronym stands for French Advances in Science and Technology. FAST is a newsletter, sent out by the French embassy, to provide "information to Americans on the achievements of France in high technology." Sad enough to see the French abasing (from the Middle French *abaisser*) themselves in front of Anglophones in such an unsubtle and shameless fashion. What do they care what Americans think? But even worse are the headlines "Wide Band Switching," "Cat Leukemia Vaccine" and "Robot Trains Come to O'Hare"—this from the country of Proust.

The French are inventing dental prostheses and new machines for kidney stone therapy; they make smart cards, the Concorde, the Ariane satellite launcher, and the Exocet missiles that were such a hit in the Falklands. They're taking their place in international high technology, and in the process they're ruining one of my favorite theories of eternal life. In this theory, which I don't think is associated with any established religion (I heard it not in a church but in a bar), the dead not only rise again but they keep their nationalities. The difference between heaven and hell is in the jobs the different national groups are assigned. In heaven, the English are the police, the Germans the engineers, and the French the cooks. In hell, the English are the cooks, the French the engineers, and the Germans the police.

FAST also reported that the French had recently built a new $500 million Science and Industry Museum at La Villette in northeast Paris. Here, in one location, was a chance to see what science had really done to the French, and vice versa. I've never shied from duty. In this I hold with Mother Teresa. I saw her on television the other day and she said that we should accept God's will. Took the words right out of my mouth. She said, to be precise, that if He wants you poor and on the street, take it with a smile. If He wants you in a palace, that's O.K. too, as long as He puts you there and you don't put yourself there. Well, He doesn't speak to me directly, at least not in English, but I got the point. He wanted me in Paris.

Since this is a science magazine I won't talk about the salmon tartare, the wild duck grilled over a wood fire, the stew of wild boar, the *tarte Tatin,* the chocolate mousse dusted with cinnamon, or the Volnay that was ordered so that its hint of raspberry would complement a peppery breast of duck, but which also seemed to go quite well with the venison. Let's just say that a country that can cook like that has no business spending its time on cat leukemia. And a man who has the chance to eat like that has no business using up valuable meal time at a science museum. You only live once, and when you die, you have no idea who the cooks will be.

There was, however, the issue of how Mother Teresa, and

the IRS, would view my trip if I didn't go to the science museum. I went, sacrificing my chance to go with my family to a place that makes its hot chocolate by melting chocolate bars. I heard later that the resulting liquid was so thick it began to harden in the cup if you dawdled. One member of my family complained that it was so rich he could only drink two cups. As for my meal at the museum, I had a pasty *pâté de foie gras* spread on cold toast. Worse yet, at one snack bar the *café express,* thick and black, was served in plastic cups. An equivalent break with American tradition would be to hawk draft beer at Yankee Stadium in teeny china demitasses. At least the wine was served in glass glasses, and, to my satisfaction, cost less than a cup of orange soda.

There were also exhibits, some of them excellent. I particularly liked the fish hatchery, which I took to be about the history and philosophy of cooking, taking trout *meunière* beyond proximate causes to its ultimate origins. There was a display of hydroponics, where salad was being grown. And this theme was picked up in a space station display. At this exhibit, a voice reminiscent of Brillat-Savarin spoke to anyone who had rented headphones. It explained how certain metal alloys could be made in weightless space and not on earth. On earth, it said, because the molten metals are of different densities, they would separate, "rather like an Italian salad dressing, in which the oil always comes to the top." Gastronomy was not dead.

"The Plastic Years," as the exhibit was called, reminded me of another national talent—for couture. The heat-retaining qualities of certain plastics were suggested by a mannequin standing on a pedestal. She was holding ski poles and had on bright red plastic ski boots, a white knit cap, a red bikini bottom but no top, and a transparent plastic coat and pants. The point here was that this outfit, of some kind of polyethylene, reflected the *infrarouge* while being

95 percent transparent to the ultraviolet—or tanning—rays. Not only that, but you could see right through it. The material hadn't been developed for semi-nude skiing, however, but for agriculture. You see, "this film permits to realize very performing greenhouses." (It's a good thing I'm bilingual.)

By the time I'd seen the skiing mannequin, I was beginning to warm to the idea of the French being involved in science, or at least in science museums. Perhaps technology was only a spice that brought out the essential flavor of the French character, like thyme on a perfectly cooked rack of lamb. It was thus in a generous, Francophilic frame of mind that I entered "Odorama."

As the name may or may not suggest, the purpose of Odorama was to demonstrate communication through smell. I passed through double doors, designed to seal in odors, and found myself in a small, dimly lit chamber with a lot of French people. On one wall was a screen and a panel to allow viewers, who in this case were also smellers, to select short film segments. First we watched children trying to burn a fox out of its den. When they lit the brand, we smelled fire. When the camera brought us inside the fox den—we smelled that. Next, a boy put bubble gum in his mouth and proceeded to walk to a barn with a friend. We experienced, in turn, the intense and unmistakable smells of bubble gum, hay (or maybe straw; I guess it wasn't that unmistakable), and cow manure.

The next bit of film was of another order altogether. It was from the movie *Swann in Love,* with Jeremy Irons. I want to warn parents who may pass this book on to their children, or borrow it from them, that this is a *French* museum we're talking about, and that *love,* in France, always seems to involve some kind of sex. Now, if you're willing to go on, place yourself, with me, in Odorama, and watch the screen:

Jeremy Irons, as Swann, is sitting alone in an elegant horse-drawn carriage, with an air of melancholy. A memory of another carriage ride thrusts itself into his consciousness. He's in the same carriage, but not alone. Next to him is a lovely woman dressed in an elegant gown with a precipitous and alluring décolletage. The allure has its effect on Swann as well as us. He's drawn to an orchid corsage anchored at the midpoint of the neckline. Gazing into the woman's eyes, he tucks the orchid into the gown. Then he bows his head to the orchid and inhales deeply. Then he gets a bit carried away—with style, to be sure, as befits a French gentleman. But soon enough the gown slips, as I've always suspected such gowns were wont to do, and it's no longer the orchid, but his lover's unclothed, unrestrained breasts (with the orchid in there somewhere) that Swann is kissing and caressing with abandon.

Meanwhile, back in Odorama, every man in the room is leaning toward the screen, inhaling deeply. I'm surprised we all didn't hyperventilate. Eventually the aroma of the orchid filled the room, Swann's recollection, and the film clip, ended, and we all straightened up, exhaling now, with sighs, as we realized that not only was this just a memory, it wasn't even our memory.

I'd always felt that passion was lacking from science, particularly from the average museum of science and industry. And yet here it was in Odorama, in full flower. What could one say, except perhaps *Vive la France*? Certainly, this was conclusive proof that my fear for the French had been misplaced. When it came to muting the French character, technology never had a chance. Indeed, I think people will soon be talking about French science with the same kind of interest and appreciation that used to be reserved for French postcards.

You see, we do need the French. In this country scientists may do research on smell and memory and sexual desire, but you would need a Ph.D. to figure out that any of this had anything to do with the sort of feelings that might make you try to inhale a movie screen in a public place. For that you need the French, because you're not going to get this version of Odorama in St. Paul, Minnesota, although there's a great science museum there. I don't think it will play in New York, either, at the American Museum of Natural History, or in Washington, D.C., at the Smithsonian. Kids go to these places, American kids. And in America we don't show heavy petting to kids, not in a museum, not with characters from Proust. We say, as Marie Antionette wouldn't have said: Let them look at minerals and gems. Later, when they're older, when they're ready, they can go to Paris— for the food, and the art, and the science.

The Fear of Cod

T wo stories:

    Somewhere in Oregon, young salmon are swimming in a Plexiglas corral set within a larger tank. The larger tank is filled with ling cod that are constantly slamming their noses against the walls of the corral in a desperate attempt to eat the salmon. The point is to train the salmon to be afraid. I'm sure it works. If I were an itty bitty fish and the giant maw of death kept crashing into Plexiglas an inch away from me, I would be even more fearful than I am now, as a full-sized human being, which is pretty fearful considering that there are no giant fish trying to eat me . . . most of the time.

Meanwhile, A. M. Rosenthal, former executive editor of *The New York Times,* the most powerful newspaper in the country, if not the world, is visiting the Galápagos Archipelago. He discovers that the marine iguanas, sea lions, and birds aren't scared of him. Presumably, it's all but inconceivable for the former executive editor of *The New York Times* to encounter a creature he doesn't frighten. Rosenthal is surprised and charmed, and reports the experience in an Op Ed page column he now writes, along with his conclusion that the absence of predation (what we in journalism call "editing") has allowed these creatures to exist in some ideal, pristine condition. He writes, "The absence of fear is the best and the original state, a thought to hug."

I call this the Huggums* hypothesis, and I put it together with the salmon story because each, in its way, suggests a meditation about fear, albeit different meditations.

First, Huggums:

The islands that inspired Rosenthal were discovered in 1535 by Spaniards and shortly thereafter became a prime destination for nature cruises. First it was the H.M.S. *Beagle,* then the American Museum of Natural History and Inca Floats vacations. Everybody who goes to Galápagos writes about them—Darwin, Annie Dillard, Kurt Vonnegut, Rosenthal, Irenaus Eibl-Eibesfeldt (I'll get to him later). Many of these writers have noted an absence of fear in some of the animals.

Darwin, for instance, found the birds quite tame, but the iguanas were afraid of him and ran away and hid in crevices. No wonder, considering the sort of things Darwin did to lizards. In *The Voyage of the Beagle* he writes, with no sense of shame that I can see, about how he confirmed that these animals, though adapted to the water, still clung for safety to the land: "I threw one several times as far as I could, into a deep pool left by the returning tide; but it invariably returned in a direct line to the spot where I stood." Darwin continues, "Perhaps this singular piece of apparent stupidity may be accounted for by the circumstance that this reptile has no enemy whatever on shore, whereas at sea it must often fall a prey to the numerous sharks." In other words, the iguanas were saying to them-

---

*Huggums is the name of a very cute doll, which really is worth hugging, and bears no relationship to any theory about fear, real or imagined.

selves: This guy Darwin may throw me around a bit but the sharks are going to *kill* me.

There are other indications that the idyl Rosenthal saw is really a kind of show put on for the tourists, that beneath the Edenlike surface lies the sad squalor of island animal life. The boobies, in particular, suffer tremendously. They're always being mugged by the frigate birds, as Eibl-Eibesfeldt documented in his book *Galápagos*. As soon as a booby got a fish, "the hovering, alert frigate birds, which had been keeping a watch on the booby, would swoop down on him and jab their beaks into his back and head." Eibl-Eibesfeldt also observed territorial posturing and ritual battles for turf and females among both sea lions and iguanas.

Of course it's possible the boobies experienced not fear but only annoyance when they were assaulted and robbed of their dinners. In any case, we can't dispute the fact that whatever else they fear, the sea lions, iguanas, and birds weren't afraid of Rosenthal. And that's enough to evince in him feelings of peace and good will. However, the Huggums hypothesis, as I read it, goes deeper. Its fundamental point seems to be that lack of fear is the "original" state. Rosenthal talks about this lack of fear being "what was meant to be." Before I explain why I think this is complete nonsense, I want to emphasize that I am no more of a scientist than Rosenthal is. In this sense we're both iguanas who may be picked up by our tails and flung out to sea by some biologist who really knows what he's talking about. What's more, Rosenthal is the dominant iguana. Still, if you don't challenge these guys once in a while you never get any females.

The only original state I can think of where fear was absent is Eden. There, it's true, the lion didn't eat the lamb. But let's be frank. Recent studies of the diet of lions have shown that as a non-miracle-based ecosystem, the Garden of Eden doesn't work. On the planet earth, for which we have better data, the situation is different. It seems pretty likely that the origin of fear has something to do with this business of getting eaten. And, in the history of life, things started eating other things almost as soon as there was anything worth eating. True, the blue-green algae weren't munched on for a billion or so years, and I would go so far as to say they never experienced fear. But they never experienced anything. By the time consciousness emerged, by the time it was possible for an organism to be aware of its environment, the world was already rife with predators, and with the fear they tend to evoke. In other words, there was no Eden, fear-wise. For most of us (poisonous caterpillars excepted for a variety of reasons) life is, always has been, and is *meant* to be scary.

I do have some actual scientific support for this idea. This is where the salmon come in. I talked to the head fish-frightener, Bori Olla, a professor of oceanography at Oregon State University's Hatfield Marine Science Center. Though it's hard for me to believe, Olla came to study salmon even though he isn't a fly fisherman. He got involved with them because he wondered how a young fish raised in a hatchery, where there's neither joy nor fear, would cope when it was released into the ocean, a place full of all sorts of emotions. (Remember the "oceanic" feeling.)

This is my phrasing. What Olla actually said was that a hatchery is a "deprived" environment, and the fish that come out of it are "naïve." He considered that these fish would face (as do all living things) two fundamental tasks in their struggle to survive—eating and not being eaten. If they couldn't catch prey they would starve, and if they were eaten . . . well, there's no need to finish that sentence. Olla wanted to know how naïveté would affect the salmon's ability to accomplish these tasks. He discovered at the outset that no matter how naïve a fish was, it quickly learned to recognize shrimp and smaller fish as food and to catch them. Avoiding being eaten turned out to be more of a problem.

First, Olla put twenty or thirty small salmon in a big tank with a ling cod, a voracious, not at all naïve predator. If these hatchery salmon were Mary Poppins, the ling cod was Mack the Knife. It went right to work and within an hour it ate half the salmon. The survivors, after having had a few lay days, were put back in the tank with the ling cod once again. They now fared much better than their naïve conspecifics. But it wasn't clear why. One possibility was that they were faster, quicker, and smarter to begin with, and that's why they survived the first time. Another was that they had learned something, namely that if there was one thing in life they should do to keep on living it was to stay the hell away from ling cod.

To see if the training had had an effect apart from cod-induced survival of the fittest, Olla introduced the corral, the Plexiglas barrier in which young salmon could be frightened but not eaten. Olla tried, in other words, to put "the fear of cod" into them.* It worked, not as well as being right in the tank with the predator, but it worked. When they faced the ling cod, trained salmon fared better than naïve salmon.

You could take this as simple confirmation of what Rosenthal was talking about. These salmon existed in a kind of dull idyl in the hatchery, and learned to be afraid when there was something to be afraid of. But Olla thinks the fish didn't actually learn fear, in the way a dog might learn not to do stuff on the rug that would get it pummeled with a rolled-up newspaper, but that the appearance of the ling cod awakened in the salmon a basic, genetic, life-enhancing capacity for complete and utter terror. Olla said that when

*This pun is the work of Ed Curtin of the department of information at Oregon State University.

hatchery fish are released into a natural environment, they seem at first unaware, dull, lacking a sense of themselves and the world around them. They're missing not only fear, but also territorial sense and other capacities that make a fish a fish. One might almost say they're not fully alive, that, in a fundamental way that has to do with their very identity, they're asleep. The ling cod wakes them right up.

And this is the real point. Fear is a part of being alive, a part of who we are, whether we're salmon or people. It isn't some sad addition to what used to be a pleasant existence. Some people even seek it out by riding roller coasters, rock climbing, spelunking, and hang-gliding. They're not grabbing gusto, they're grabbing fear. And the reason is that they don't want to feel that they're living in a hatchery. I guess I'm one of the lucky ones. I've never needed artificial stimulation to be frightened out of my wits. I experience a lot of fear just sitting at my desk. The list of things I'm afraid of (giant fish, A. M. Rosenthal, fear itself) is incredibly long, and what I take this to mean is that I'm living life to its fullest, every minute of every day.

Playing Doctor

Earlier this week I killed three emergency room patients in the space of about fifteen minutes. I sent two into respiratory arrest and one into cardiac arrest. (This is known among trauma interns as a hat trick.) Of course it's possible that they didn't die. For all I know, after I blew it some other doctor called a Code Blue or a Code Purple (idiot in the emergency room) and saved Eugene Wilson, 21, motorcycle accident victim, or Donnie Brooks, 20, the victim of a stabbing incident outside the River Tavern, which, judging from Donnie's appearance before I let his heart stop, isn't the kind of place I'm ever going for a drink or a stabbing incident.

I know this sounds bizarre, and the truth is there weren't really three different patients. The truth is I killed Donnie Brooks twice. I was able to do this for two reasons. First, I'm not a doctor. Second, I did it on DxTer, a combination of video and computer technology that's used to train medical students the way flight simulators help train pilots— a kind of "St. Elsewhere" simulator.

DxTer is part of the current interaction bonanza. By interaction I don't mean conversation or psychotherapy or being mugged—the classic forms of interaction. Those all involve other people. The boom now is in interacting with television of one sort or another, often for educational purposes. Usually this means you affect what's happening on the screen, although sometimes the screen affects you. For example, one company is making a set of remote-controlled robots for children's toys. The kid controls the good guys but the bad guys are controlled by inaudible signals from a television program on which they also appear. I suspect the show will also turn on your vacuum and drive your car if you don't watch out.

Moving up in age there's the interactive video courtroom at Harvard Law School, which allows students to act as counsel in a videotaped trial that appears on a computer monitor. The students use a computer keyboard to raise objections and cite precedents. The system is limited, of course. There's no provision for the force of oratory, or for bribes. There's also a law enforcement simulator. In one of this system's training dramas a life-sized woman on a large screen reaches into her pocketbook for either a driver's license or a gun. The law enforcer has to decide whether or not to blow her away. If he shoots, bullet holes appear on the screen and the action stops. I suppose this is what today's children think happens when you die. You're going along at regular speed in the movie of life, and the next thing you know—freeze frame! Or else you suffer a prolonged period of slow motion before the end.

The goal of all this educational interaction is to make better doctors, cops, and lawyers by letting them try out their knowledge in something more demanding than a written exam before we give them accident victims, live ammunition, and real money. I'm sure the simulators do what they're meant to, and that they'll all benefit society tremendously. But that isn't what interested me. What interested me was that all this was taking place on TV. Officially, it's video, on a computer monitor. But really, it's TV. It's happening on the same little screen that entranced us in childhood, the same screen that held me in thrall throughout my years of higher education, thus derailing my medical, legal, and law enforcement careers and forcing me to seek a less reputable form of employment. It's on the same screen on which I first saw "Sky King" (a flight simulator) and "Rawhide" (a masculinity simulator). It was on this screen that I watched "Gunsmoke," "Dragnet," "Maverick," "Perry Mason," "Dr. Kildare," and "Car 54, Where Are You?"

So when I read about the simulators I said to myself, Ooh! Ooh! Ooh! I got an idea! The idea, naturally, was to try out a video simulator, not because I wanted to learn to be a better doctor or cop, but because after years of watching TV shows I wanted to be *in* one.* The only question was which one. I didn't have much interest in lawyer shows. The law has never looked like a fun profession to me, except for the chance to hang out with Joyce Davenport. And while there was a certain appeal to the unrestrained violence of police dramas, as I understand it when the people on the cop simulator die, they don't bleed. On DxTer, they bleed.

Hospital drama definitely had the most appeal. It had great characters. I would be free to imagine myself as Ben Casey or Marcus Welby if I happened to be feeling grandiose, as the anxious and unpleasant Ehrlich on "St. Elsewhere" if I happened to be in a period of low self-esteem, or, most likely, as my all-time favorite television doctor, Hawkeye Pierce of the 4077th M*A*S*H. And in hospital shows, even if the characters are weak, or the plot fails to click, there's other interesting stuff—you know, blood gases, X-rays, ampicillin. The choice was clear.

---

*I'm sure all the people who use these things imagine themselves in TV or the movies. You can't tell me that rookie law enforcers, when the sergeant isn't around, don't give the dame with the pocketbook a crooked smile and say, just before they squeeze the trigger, "Go ahead, lady, make my day!"

I flew to San Diego to meet DxTer and talk to its creator, a doctor. His name is David Allan. He's the head of Intelligent Images, Inc. and he showed me the ropes on his brainchild, which is already in use in several medical schools and is being marketed by IBM. DxTer consists of a computer program, the computer itself, a video disc player, and a collection of video discs on which eight different cases are recorded, including stab wounds (Donnie Brooks), shotgun wounds (Victor Mercedes), chest trauma (Eugene Wilson), and diarrhea and vomiting (John Dircus). All are based on real medical cases and have been staged as short hospital dramas with actors and excellent special effects. This means that everything looks real. When you see Eugene Wilson gag and sputter while the trauma team intubates him (this involves thrusting what looks to be a plastic spout down his throat), it makes you incredibly glad you didn't get into medical school.

The cases, or episodes, are stored on video discs because a computer can immediately call up to a monitor any part of a disc; there's none of the rewinding and fast-forwarding required with tape. DxTer's computer program is the heart of the system. It determines what section of the video disc to show and tells the computer when to switch from the emergency room drama to a printed menu that offers various courses of action for the viewer to take. The screen is touch sensitive, so all you have to do is push "Examine" or "Diagnostics" or "Medicine/Therapeutics" to be offered a new list of possible medications or tests. All the arcana of medicine are at your fingertips. The program specifies a response to every possible decision a viewer can make, whether it's to flash images of the intubation the student has just ordered, or to suggest, politely, that when a man has uncontrollable diarrhea, it's no time to X-ray his foot.

I watched Allan whiz through the treatment of Eugene

Wilson once to pick up a few pointers before I took over. When I did, things happened all too fast, particularly for the patient.

Suddenly Eugene Wilson was wheeled into the emergency room, bleeding, bandaged, and wearing a pair of huge inflatable trousers to squeeze his legs and push the blood up into his thorax (do I sound like a doctor or what?). The paramedic said something like, "You think he's bad, you should've seen the bike." The head nurse looked at Eugene Wilson's shock pants and said, "Anything under the trousers?" I laughed. Nobody else did. Clearly this wasn't Korea.

The trauma team set to work on him. Then the screen switched to the menu, putting the victim in my hands. "Examine" was an obvious first choice. I pushed it and the team began to examine him. He wasn't in good shape. He reminded me of the old song, "There was blood on the saddle and blood on the ground, and a great big puddle of blood all around." I had him intubated (why not?), took a bunch of X-rays, and prescribed all the antibiotics I could get my hands on. DxTer implemented my orders. Occasionally DxTer acted on its own. It flashed the message, "The patient must be ventilated immediately," and then proceeded to have the team do it. At other times cryptic messages appeared on the screen, like "Capillary refill is delayed." What? I ordered a few more tests, some fluids, and tried to remember an episode of "M*A*S*H" that might help me. I was floundering; it was becoming obvious to me why people go to medical school.

The patient took a downhill turn. I ordered a consultant. He was on his way. I ordered a chaplain. Suddenly there was Eugene Wilson on the screen again, looking even worse. The nurse turned to me, and said, in what I thought was a rather accusatory tone, that the patient wasn't breathing. Why do you think I called the chaplain? I thought to myself.

Now I was starting to get irritated. I'd already cleared this guy's airway and intubated him. What was his problem? I decided to give him a proctoscopy. I figured that ought to wake him up. DxTer demurred. The proctoscopy wasn't indicated. The next thing I knew respiratory arrest ensued, I was booted out of the program and the emergency room, and DxTer suggested I try again. Well, if you ask me, if we'd given him the proctoscopy Eugene Wilson would be alive and walking around today, although perhaps not comfortably.

This wasn't the way the experts saw it. DxTer will give you an expert analysis of how you botched a case, if you ask for it. The experts claimed that Eugene Wilson had had a tension pneumothorax and that I should have put in a chest tube. Experts. They think they're so smart. I moved on to Donnie Brooks, to try my hand at stab wounds. The first time I neglected to clear his airway and he choked on his tongue, and the second time he didn't get enough fluids so he went into shock and his heart stopped. On that last try, I got a cost analysis of my work. The so-called experts may have saved Donnie Brooks, but they cost the hospital $1,852. I did much better. I got us out for under a grand— $842—not bad for a River Tavern knifing, I thought.

I certainly learned a few things from DxTer. I now know that if anybody at a party comes down with a tension pneumothorax, what you do is slap a chest tube into him. I learned what shock trousers are. I expect to see David Bowie or David Byrne—some David anyway—wearing them soon. And if I picked up this much, DxTer has got to be even more valuable to medical students, who actually know what a tension pneumothorax is, I hope.

DxTer isn't great TV, however. The thing is that as a doctor, I'm far less interesting, even to myself, than Alan Alda. And I realized as I dealt with the incessant demands of the patients—always bleeding, always needing fluids, or ventilation, or some tube or other—that I didn't need this grief. It wasn't my job. I've got problems of my own. And I get more than enough interaction in everyday life. What I want from television is a plot that just chugs along on its own, with no effort on my part. I want to watch the shows, not write them.

On the other hand, I'm all for the proliferation of training simulators for situations in which it's dangerous to jump right into the real thing, like marriage. There could be a marriage simulator with a choice of scenarios: Infidelity, Overdrawn Checkbook, Let's Talk About Me. A headwaiter simulator would also be useful. It could be wired to deliver a severe dose of humility if you couldn't read the menu and an electric shock if the tip were too low. And I'm sure we'll see others as well: maybe one for the robbers so they can decide whether or not to shoot the cop, perhaps even one for emergency room patients. I can see it now: first there'll be the roof of the ambulance, then the doctor looming over you, and then you'll get a menu of choices: "Do you want to a) Choke on your tongue, b) Go into cardiac arrest, or c) Open your eyes wide and say, 'Thanks, Doc. Say, whose pants are these anyway?' "

That Little Piggy

Never underestimate the power of the press. If at first you don't succeed, try, try again. If the trout aren't rising, fish a gold ribbed hare's ear nymph (weighted). These are lessons I learned as an investigative reporter during my time on the trail of the micro pig. The micro pig (there are actually many individual micro pigs) is a very small, very friendly variety of swine bred at the Miniature Swine Research Laboratory at Colorado State University in Fort Collins. It isn't so miniature, but, as pigs go, it's pretty small. Micro pigs weigh sixty to one hundred pounds and are bred to serve as laboratory animals for research on everything from diabetes to nutrition and exercise.

This does not, I suppose, sound like a subject for an investigative reporter. But it isn't that easy to get to see a micro pig—at least it didn't used to be. My first attempt to see the pigs (which I planned to combine with a trout fishing trip) was foiled by the forces of darkness. Someone at Colorado State had decided I might make these pigs sound too cute and that this might cause problems with the animal rights movement. My request to see the pig was refused.

I wrote a column pointing out that this sort of thing wasn't in the spirit of the Constitution (or the bylaws of Trout Unlimited, for that matter). I said a few things about the "Pork Curtain." I asked readers to mount a letter campaign. The response was . . . well, there was a response. I know for a fact that at least four people—Janice Thompson, Mrs. Kevin Martin, Tyler Suchman, and Andy Markley—struck epistolary blows on my behalf. All wrote in the spirit that Markley captured so well in his closing lines: "We want to see the micro pig! We have a *right* to see the micro pig! . . . Show us the micro pig!"

By the standards of the eighties I had clearly created a movement, if not a revolution. Shortly after the column came out, and the four letters began pouring in, Jim Bolick of the Colorado State Office of University Communications called me, apologized, and invited me out to see the pig. I knew then how Vladimir Ilyich Lenin must have felt in 1917 as he planned his trip to Moscow. Of course, the Czar never apologized to Lenin, which is why things got out of hand. It takes a big university to admit it's wrong, and CSU is a big university. In fact, Bolick not only invited me to go see the pig, but he also offered to take me fishing to make amends, since I had complained bitterly about missing out on all the giant trout I figured lived in Colorado along with those tiny pigs. After giving his offer some thought, I agreed to go see the pigs, since I felt I owed something to movement members, to the profession of journalism, and, if you want to think about it, to CSU. However, I declined the fishing offer.

I had several reasons. For one thing it occurred to me that after my first article perhaps what CSU had in mind wasn't a fishing trip but an accidental drowning. If not, well, journalists are supposed to be objective. One of the things they teach you in journalism school is that you should try not to be influenced by the benevolence of your sources into being overly fair to them. I felt I had a record to uphold. I've never been accused of fairness before, and I didn't want to start now.

So, I took the high road. No, thank you, I said to Bolick, I'll go fishing on my own. I emphasized the size and strength of my journalistic integrity (macro, very macro), and told him I would never use my position as a journalist to gain personal favors. However, I said, if by chance there were somebody at CSU in fisheries biology that I could talk to, I might like to get in a little extra work (I'm a glutton for work). Could he, I asked, set up an interview for me on the subject of current trout research. You know: Where are the big ones? What are they taking? Do journalists get to fish in the hatchery ponds? Yes, Bolick said, he thought there might be somebody available for the sort of interview I had in mind.

By this circuitous path I came, after many months, to the swine lab. And now it's my turn to apologize. In my earlier pig foray, I had had the questionable taste to wonder whether the pig was really as small as CSU claimed. That wasn't nice. Having now seen roughly 150 micro pigs I can verify that Linda Panepinto, who bred them, has indeed succeeded in developing small pigs. I find it hard to convey, for those of you who have no farm experience, what a sixty-pound adult pig looks like. The best I can do is to say that it's about the size of a beagle—an incredibly obese beagle. Since it's a pig, however, it carries the weight well. As to the whole worrisome question of cuteness in a laboratory animal, you might say these pigs are cute, but ugly cute, not cute cute. For instance, big pigs have a lot of hair. Some of these micro pigs are almost completely bald, others have scraggly wisps of hair. Remember Ho Chi Minh? Remember his beard? Well, the micro pigs that do have hair have got Ho Chi Minh's beard over their entire bodies.

The pigs also have wattles. I wouldn't mention this except for my enormous journalistic integrity, because the subject is probably embarrassing to the pigs. These aren't cute wattles. These are pendulous fleshy growths, one on each side of the head, in the general area of the jowls. Not that this makes the pigs entirely unappealing. Wattles of this sort wouldn't look good on a poodle, but on a pig they're acceptable.

I had also made some earlier noise about small pigs losing the dignity that went along with size. I no longer recall where I got the idea that pigs had dignity, but whether or not they do, these pigs do retain, by virtue of their physical characteristics, an inherent, undeniable pigness. There's something about them—some of them, anyway—that says, in a loud grunt, "I am swine."

Once I had seen the pigs the next question was as obvious to me as it would have been to any investigative reporter: Where can I catch a trout the size of a micro pig? Robert J. Behnke, professor of fisheries biology at CSU, fly fisherman, columnist for *Trout* (the magazine of Trout Unlimited), had seen me coming. In fact, I had telephoned him first. He knew I was interested in big fish. So, shortly after I sat down in his office he handed me a newspaper clipping about a trout in China that had eaten at least fifty horses and a dozen people. Normally I wouldn't believe this kind of story, particularly since it was published in one of those tabloids that usually feature items on the children of alien visitors born to earth virgins. (Who are they kidding? Virgins?) But in fly fishing (a science of sorts) it's well known that trout become fixated on the most plentiful food around (in the Chinese river, obviously horses). With smaller trout it's usually some kind of insect that's in the process of hatching—that is, going from its aquatic, larval stage to its aerial, adult stage. You try to present an artificial fly that looks like the food the trout has its mind on, and the process is called matching the hatch. This would be one hell of a trick for a fish that ate horses. But you would have to use artificials. Putting a real horse on a hook would be live-bait fishing at its most grotesque.

Behnke explained that the story was false (I knew that), although it did have a grain of truth. There are indeed big trout in China, the famed Huchin trout, which sometimes grow to a size of one hundred pounds. Not only that, but these fish also exist in parts of the Soviet Union, and, said Behnke, Trout Unlimited was in the process of organizing a delegation of American anglers to go to the Soviet Union to meet Russian anglers and do a little fishing. I assume this venture is being undertaken in the interests of international diplomacy, just as my trip to Colorado was in the interest of science and First Amendment freedoms.

Having realized that the real giant trout weren't in Colorado but in Asia, I tried not to lose heart. I went fishing, the day after I saw the pigs, in the Cache La Poudre (hide the powder) River, about an hour from Fort Collins. As to what happened that day, it all depends on what kind of a writer you are—what you put in, what you leave out, how you say it. When it comes time to talk fishing, those of us who take shelter under the First Amendment are slipperier than greased micro pigs.

For instance, I could tell you about the sun glinting off the gin-clear stream as it tumbled around rocks, and the rainbow trout flashing white under the surface as they turned in the current to snatch nymphs. I could speak of the fish rising to sip emerging midges (tiny flies—there's no end of smallness in this world) at the surface. There's a lot of this talk in fishing magazines. It's angling erotica. There was a bit of it in *Harper's* while I was on this trip. The piece was by Thomas McGuane and although it was elegant and witty, not at all the bathetic, trout-benighted ode one is apt to find in the other fishing magazines, McGuane did occasionally drift into an elegiac, almost erotic mode, lavishing on the river and its trout the kind of descriptive prose a man who didn't fish might reserve for the body of a beautiful woman.

There is, of course, another side to fishing, and women. Lust, of all sorts, often goes unrequited. Lovers get jilted, fishermen get skunked. There's bedroom slapstick, and there are the numerous accounts of fishermen who hook themselves, fall down, get wet, and drown. In this vein I could tell you about the leak in my waders, or how I fell headlong and cracked both my shins and the handle of my new fly rod. I could report the sign I saw in the first spot I fished— BIGHORN SHEEP LUNGWORM TREATMENT AREA. DO NOT APPROACH SHEEP. I could tell you how, after an hour or two of seeing, but not catching, fish I decided that fly fishing was an idiotic pastime designed for effete imitators of British royalty, and that what I really needed was a few of those lungworms for bait.

Or, if I were a good honest reporter I suppose I would stick to the facts, which are that in the end I found a better spot and I caught seven rainbow trout from eight to ten inches long using a size 14 weighted hare's ear nymph fished upstream with a strike indicator. As to how to describe the trout, they weren't macro trout certainly, but they weren't micro trout either. I've caught a lot of five-inch brookies in my day—*those* are micro trout. You could call these mini trout. Or, if you were McGuane, who described his brown trout as "short, thick, buttery"—you might call these fish, oh, I don't know, shiny maybe, or Christmas tinsel bright. Obviously, I'm not McGuane. And, although I hate to pick on the poor pig people again, I think it's cruel to saddle animals with computer terminology.

Small. That's what I'd call the trout I caught. Small.

Night of the
Living Optimists

ow's this for a horror movie? A psychologist (scary
already, isn't it?) develops a questionnaire to tell
who is an optimist and who is a pessimist. The test
is designed to benefit humanity, just like the new genetically engineered bacteria. But it gets loose and falls into
the hands of the insurance industry, which uses it to develop
a task force of superoptimists, men and women who will
never admit defeat, who always have a cute story ready,
who believe deeply in the product, whatever it is.

Cut to a shapely young woman taking a shower. The
doorbell rings. She puts on a towel and goes to the screen-
door. The background music is all funny chords, sevenths
and ninths in minor keys. At the door is a smiling man
with an unnatural resemblance (physically and philosoph-
ically) to Ronald Reagan. Crescendo. He is selling life insurance. Screams. The towel slides to the floor. He keeps
on selling life insurance.

Sure, you say, it would make a scary movie, but it's like
Alien, where the thing with the teeth pops out of the guy's
chest. It's not real; it's just some writer's nightmare. It could
never happen in *my* town. Well, if that's what you think,
you might want to pick up Volume 50 of the *Journal of
Personality and Social Psychology* and read "Explanatory

Style as a Predictor of Productivity and Quitting Among Life Insurance Sales Agents," by Martin E. P. Seligman and Peter Schulman of the University of Pennsylvania. There *is* such a test and it *has* fallen into the hands of the insurance industry.

What Seligman and Schulman did was to use the test (Attributional Style Questionnaire) on insurance sales agents for Metropolitan Life Insurance Company in Pennsylvania to find out whether they had what Seligman calls optimistic or pessimistic explanatory styles. The optimists sold more insurance and kept their jobs longer. The pessimists sold less, and presumably seeing that there was no hope of improvement for them and that it was a horrible job anyway, quit earlier. (Not that one can assume they thought they were going on to anything better.) And there is indeed a task force, although not exactly as I described it. The real one is composed of people who failed the standard industry test used to pick insurance sales agents, but whom the ASQ pegged as optimists. At last report they were doing better than anyone else.

This is only one piece of research. There are quite a few others. In fact, optimism is very big these days, because it seems that it's good for you. There's evidence that in addition to selling more insurance, optimists live longer and their immune systems work better. Not only that, but some people, Seligman among them, have suggested that one may be able to change from pessimism to optimism. I wonder. How could a die-hard pessimist muster the initial optimism necessary even to *try* to become more optimistic? Wouldn't he say: "Oh, what's the point? I'd never be able to change. I'm just a pessimist at heart"?

The optimism business seems to have begun with giving electric shocks to dogs. These experiments, which are quite famous, showed that if the shocks were inescapable the animals realized this and just gave up. This so-called learned helplessness seemed to Seligman to be somewhat similar to human depression. I guess the idea was that some humans, for whatever reasons, seemed to think of life as a series of inescapable electric shocks (I can't imagine why) and were lying down in their cages and giving up.

This perception led Seligman and others to consider, over a number of years, the ways people explain bad events that happen to them, like electric shocks. In the current formulation, an optimist, when confronted with a personal failure like not selling an insurance policy, interprets this in external, specific, unstable (temporary) terms. He says, It's not my fault; this particular customer was a jerk. The next one will be a sucker for sure. The pessimist interprets failure in internal, global, stable (permanent) terms. He says, It's my fault, I'm an idiot, I've always been an idiot, I'll always be an idiot. Oh God, there's just no point in living. Of course, there are variations, cases that are hard to categorize. For instance, somebody could come up with an internal, global, temporary explanation: I'm an idiot; there's no point in living *today*. Pessimists, since they tend to blame themselves for everything, are more likely to get depressed and give up when bad things happen to them. Optimists just whistle a happy tune and blame someone else.

What strikes me about the interest in optimism is that it is, in essence, a reemergence of the Pelagian heresy, which I thought had been taken care of once and for all at the Council of Carthage in 418. But these things never die. Other people, smarter than me (some of them anyway), have pointed out that science continually plunders the past for its paradigms. There are only so many ideas around, and they keep resurfacing with new paint jobs. In a sense, all human intellectual endeavor is nothing but one big chop shop, disassembling, repainting, and filing the serial number off old ideas—in this case the Pelagian heresy (a 1987 Nissan 300Z) and Augustinian orthodoxy (a 1936 Rolls-Royce Phantom).

I realize that not everyone has kept up with the doctrinal twists and turns of early medieval Roman Catholicism, so let me explain. The Augustinians (following St. Augustine) pessimistically believed, and still do, that each human being is born blotched with original sin and can be redeemed only by God's grace. This view is often expressed in the liturgical lament: *mea saurus, mea saurus, mea maxima saurus.** The Pelagians saw the belief in original sin as having to do with low self-esteem. (They were way ahead of their time. Of course they lived in the fifth century, when it was hard not to be ahead of your time.) They said: Hey! I'm okay, you're okay; if we just work hard and do good we can get to heaven on our own—*whether God likes us or not.* You can imagine how this infuriated the Augustinians.

Intellectual positions become transformed over time, of course, and today the issue is not salvation, but athero-sclerosis. True Pelagianism has been transformed, not so much into optimism itself, but into the medical and psychological belief that, if you are optimistic, you can, in worldly terms, save yourself—meta-optimism.

The original Pelagians did not fare well. They were branded heretics, which was, in those days, like having the National Science Foundation cut off your funding. Obviously the early bishops, and Pope Zosimus, who confirmed and validated the verdict of the Council of Carthage, felt there was some danger to the bright, happy, Pelagian approach to life, or afterlife. If so, I've finally found some common ground with Pope Zosimus. I don't doubt that optimism does wonders for the optimists themselves. The question I have not yet seen addressed and want the answer to is this:

*Loosely, "I'm a lizard, I'm a lizard, I'm an enormous lizard."

What does optimism do for, and to, the rest of us? Instead of thinking about the salesmen, let's think about the customers.

Who in their right mind likes a cheery, cocksure life insurance salesman? You might as well talk to a Mormon missionary. I know this seems like a gratuitous use of Mormonism, but it's not. You see, the Mormons have a history of optimism. Joseph Smith himself is an example of the most monumental optimism (or *chutzpah,* perhaps, except the word seems out of place in this context). Here was a guy who thought—in the nineteenth century—that he could start a brand-new, giant religion from scratch. And he succeeded. That's the scary part about optimists: they tend to succeed. And that means that the more of them there are out there selling life insurance, the more life insurance the rest of us are going to have to buy.

One can't fail to note that Seligman did not study shoe salesmen. Presumably this is because selling shoes isn't that stressful. One doesn't need to be an optimist to sell shoes. Why? Because people need shoes. They want shoes. They like shoes—wing tips, high tops, sandals, Hush Puppies, shoes are very popular. If you want to sell them you don't have to abuse your alumni directory and send out a bunch of form letters that start, "Dear classmate: Now that you're a father and provider it's time you began thinking about shoes." No. Shoe salesmen just sit in their stores and wait. They may wonder why their lives are spent handling other people's feet, but they're at least secure in the knowledge that as long as other people have those feet, shoe salesmen (optimists and pessimists alike) will make a living.

My feeling is that if insurance were all that great, you wouldn't have to be an optimist to sell it. Maybe insurance isn't so great, not as great as shoes anyway. And maybe optimism isn't so great either. Maybe pessimists are better for the general weal. Sure they're lousy life insurance salesmen, but so what? Balanced against this small failing is a profoundly beneficial quality that all pessimists share. They tend not to *do* much. Nowadays pretty much everything that gets done is bad, so pessimists are continually performing what you might call, to fuse once and for all the Judaic and Christian religious traditions, *mitzvahs* of omission.

For example, when there's some covert action to be undertaken, say the destabilization of the Southern Hemisphere, the bigwigs are always looking for "can do" people. What if they could only find "can't do" people? "No," the can't-do's would say, when you asked them to assassinate a foreign head of state, "that will never work. The poison will fail, the gun will jam, plus, I can't keep a secret." A pessimist president, when things went wrong, would tend to blame himself, which would be refreshing. He also might try to avoid things that could go wrong in a big way, like wars. And if we could only elect a dour, pessimist Congress, I'm sure it would refuse to fund nuclear weapons. The representatives and senators would realize, as human beings themselves, that, with the exception of those of us who are actually crooked, we're a race of fools, lunatics, and insurance salesmen, and sooner or later somebody in one category or another is going to set the bombs off.

Pessimism: it's our only hope for the future.

Biblioasis International Translation Series
General Editor: Stephen Henighan

Jorge Carrión

BOOKSHOPS
A Reader's History

Translated from the Spanish by
Peter Bush

BIBLIOASIS
WINDSOR, ONTARIO

381.45
CAR

Library and Archives Canada Cataloguing in Publication

Carrión, Jorge, 1976-
[Librerías. English]
    Bookshops : a reader's history / Jorge Carrión.

(Biblioasis international translation series ; no. 21)
Translation of: Librerías.
Translated from the Spanish by Peter Bush.
Issued in print and electronic formats.
ISBN 978-1-77196-174-5 (hardcover).--ISBN 978-1-77196-175-2 (ebook)

    1. Bookstores. 2. Books and reading. 3. Authors--Books and reading.
I. Title. II. Title: Librerías. English. III. Series: Biblioasis international translation series ; no. 21

Z278.C3113 2017   070.5   C2017-901938-4
                                                            C2017-901939-2

Readied for the Press by Daniel Wells
Copy-edited by Emily Donaldson
Typeset by Chris Andrechek
Jacket Designed by Jeffrey Fisher

PRINTED AND BOUND IN THE USA

A bookshop is only an idea in time.

Carlos Pascual, "The Power of a Reader"

I do not doubt that I often happen to talk of things which are treated better in the writings of master-craftsmen, and with more authenticity. What you have here is purely an assay of my natural, not at all of my acquired, abilities. Anyone who catches me out in ignorance does me no harm: I cannot vouch to other people for my reasonings: I can scarcely vouch for them to myself and am by no means satisfied with them. If anyone is looking for knowledge let him go where such fish are to be caught: there is nothing I lay claim to less. These are my own thoughts, by which I am striving to make known not matter but me.

Michel de Montaigne, "On Books,"
translated by M. A. Screech

A man recognizes his genius only upon putting it to the test. The eaglet trembles like the young dove at the moment it first unfolds its wings and entrusts itself to a breath of air. When an author composes a first work, he does not know what it is worth, nor does the bookseller. If the bookseller pays us as he wishes, we in turn sell him what we are pleased to sell him. It is success that instructs the merchant and the man of letters.

Denis Diderot, "Letter on the Book Trade"
translated by Arthur Goldhammer

CONTENTS

Introduction inspired by a Stefan Zweig short story

The way a specific story relates to the whole of literature is similar to the way a single bookshop relates to every bookshop that exists, has existed and will ever perhaps exist. Synecdoche and analogy are the two most useful figures of speech: I shall start by talking about all bookshops from the past, present and whatever the future may hold via one story, "Mendel the Bibliophile," written in 1929 by Stefan Zweig and set in Vienna in the twilight years of empire, and will then move on to other stories that speak of readers and books in the course of a frenzied twentieth century.

Zweig does not choose a renowned Viennese café for his setting, not the Frauenhuber or the Imperial, one of the cafés that were the best academies for studying the latest fashions—as he says in *The World of Yesterday*—but an unknown café; the story starts when the narrator goes to "the outer districts of the city." He is caught in the rain and takes shelter in the first place he finds. Once seated at a table, he is struck by a feeling of familiarity. He glances at the furniture, the tables, billiard tables, chessboard and telephone box, and senses that he has been there before. He scours his memory until he remembers with a sharp jolt.

He is in the Café Gluck, where the bookseller Jakob Mendel had once sat, each and every day from 7.30 a.m. to closing time, surrounded by heaps of his catalogues and books. Mendel would peer through his spectacles memorizing his lists and data, and sway his chin and curly ringlets in a prayer-like rhythm: he had come to Vienna intending to study for the rabbinate, but antique

9

books had seduced him from that path, "so he could give himself up to idolatry in the form of the brilliant, thousand-fold polytheism of books," and thereby become the Great Mendel. Because Mendel had "a unique marvel of a memory," he was considered "a bibliographical phenomenon," "a *miraculum mundi*, a magical catalogue of all the books in the world," "a Titan":

> Behind that chalky, grubby brow, which looked as if it were overgrown by grey moss, there stood in an invisible company, as if stamped in steel, every name and title that had ever been printed on the title page of a book. Whether a work had first been published yesterday or two hundred years ago, he knew at once its exact place of publication, its publisher and the price, both new and second-hand, and at the same time he unfailingly recollected the binding, illustrations and facsimile editions of every book [. . .] he knew every plant, every micro-organism, every star in the eternally oscillating, constantly changing cosmos of the universe of books. He knew more in every field than the experts in that field, he was more knowledgeable about libraries than the librarians themselves, he knew the stocks of most firms by heart better than their owners, for all their lists and card indexes, although he had nothing at his command but the magic of memory, nothing but his incomparable faculty of recollection, which could only be truly explained and analysed by citing a hundred separate examples.

The metaphors are beautiful: his brow looks "overgrown by grey moss," the books he has memorized are species or stars and constitute a community of phantoms, a textual universe. His knowledge as an itinerant seller without a licence to open a bookshop is greater than that of any expert or librarian. His portable bookshop, with its ideal location on a table—always the same one—in the Café Gluck, is now a shrine of pilgrimage for all lovers and collectors of books, as well as all the people who could not find the bibliographical references they had been looking for via more official routes. After an unhappy experience in a library, the narrator, a young university student, is taken to the legendary café table by a friend, a guide, who reveals to him a secret place that does not appear in guidebooks or on maps, one that is only known to the initiated.

Rabbi Jacob Mendel Morgenstern, rabbi of the Great Synagogue in Wegrow. He was the son of the rabbi of Sokolow. When the Nazis first entered Wegrow, they took him to the town square, made him clean the streets, and then bayoneted him to death.

One could include "Mendel the Bibliophile" in a series of contemporary stories that focus on the relationship between memory and reading, a series that might start in 1909 with "A World of Paper" by Luigi Pirandello and finish in 1981 with *The Encyclopedia of the Dead* by Danilo Kiš, by way of Zweig's story and the three Jorge Luis Borges wrote in the middle of the last century. The old meta-book tradition reaches such a level of maturity and transcendence in the world of Borges that we are duty-bound to consider what comes before and after each of his stories as precursors and heirs. "The Library of Babel," from 1941, describes a hyper-textual universe in the form of a library hive devoid of meaning, one in which reading is almost exclusively a matter of deciphering (an apparent paradox: in Borges' story reading for pleasure is banned). Published in *Sur* four years later, "The Aleph" is about how one might read "The Library of Babel" if it were reduced to the tiniest sphere that condenses the whole of space and time. And above all, it is about the possible translation of such a reading into a poem, into language that makes of the portentous *Aleph*'s existence something *useful*. But "Funes the Memorious," from 1942, is undoubtedly the Borges story that most reminds us of Zweig's, with its protagonist who lives on the edge of Western civilisation, and who, like Mendel, is an incarnation of the genius of memory:

> Babylon, London and New York have overawed the imagination of men with their ferocious splendour; no-one in those populous towers, or upon those surging avenues, has felt the heat and the pressure of a reality as indefatigable as that which day and night converged upon the unfortunate Irineo in his humble South American farmhouse.

Like Mendel, Funes does not enjoy his amazing gifts of recall. Reading does not involve a process of unravelling plots for either of them, nor is it about investigating life patterns, understanding psychological states, abstracting, relating, thinking, experiencing fear and pleasure on their nerve-endings. Just as it is for Number 5, the robot in the film *Short Circuit*, which appeared forty-four years later, reading for them is about absorbing data, myriad labels, indexing and processing information: desire is excluded. The stories by Zweig and Borges complement each other entirely: old man and young, the total recall of books and the exhaustive recall of the world, the Library of Babel in a single brain and the Aleph in a single memory, both characters united by their poverty-stricken, peripheral status.

In "A World of Paper," Pirandello also imagines a reading scenario beset by poverty and obsession. A compulsive reader to the extent that his skin replicates the colour and texture of the paper, but deep in debt because of his habit, Balicci is going blind: "His whole world used to be there! And now he could not live there, except for that small area his memory brought back to him!" Reduced to a tactile reality, to volumes as disorganized as pieces of the Tetris, he decides to contract someone to classify his books, to bring order to his library, so that his world is "rescued from chaos." However, he subsequently feels incomplete and orphaned because he finds it impossible to read. He hires a woman reader, Tilde Pagliocchini, whose voice and intonation annoy him so much that the only solution they can find is for her to read extremely quietly—that is, silently—so that he can imagine his own, ever-diminishing reading habit from the speed at which she covers lines and pages. His whole world, re-ordered in memory.

A world that can be encompassed and shrunk thanks to the metaphor of the library, a portable library or photographic memory that can be described and mapped.

It is not merely fortuitous that the protagonist of *The Encyclopedia of the Dead* by Danilo Kiš is, in fact, a topographer. His whole life has been determined in minute detail by a kind of sect or band of anonymous scholars who, from the end of the eighteenth century, have been pursuing an encyclopedic project—parallel to that of the Enlightenment—which includes everyone in history not to be found in all the other official or public encyclopedias available for consultation in any library. The story goes on to speculate about the existence of a Nordic library where one might find the rooms—each allocated a letter of the alphabet—of *The Encyclopedia of the Dead*, one where every volume is chained to its shelf and is impossible to copy or reproduce, the object of partial readings that are immediately forgotten.

"My memory, sir, is like a garbage disposal," says Funes. Borges always speaks of failure: the three wonders he has conjured up amount to gestures in the face of death and the absurd. We know how foolish these lines are that Carlos Argentino was able to write with the inspiration of the incredible Aleph, the possession of which he irrevocably squandered. And Borges' librarian, a persistent explorer of the library's nooks and crannies, lists in his old age all the certainties and expectations humanity has gradually discarded over the centuries, affirming at the end of his report, "I know of districts where the youth prostrate themselves before books and barbarously kiss the pages, though they do not know how to make out a single letter." We find the same elegiac tone in the stories I have mentioned: Pirandello's hero goes blind, Mendel is killed, the Library of Babel loses habitués to lung disease and suicide, Beatriz Viterbo has died, the father of Borges is ill and Funes dies of congestion. The father of Kiš's narrator also disappears. What links the six stories is an individual and a world in mourning: "Memory of indescribable melancholy: walking many a night along gleaming passageways and stairs and not finding a single librarian."

So I was overcome by a kind of horror when I saw that the marble-topped table where Jakob Mendel made his oracular utterances now stood in the room as empty as a gravestone. Only now that I was older did I understand how much dies with such a man, first because anything unique is more and more valuable in a world now becoming hopelessly uniform.

Mendel's extraordinary nature, says Zweig, could only be recounted through examples. To describe the Aleph, Borges has recourse to the chaotic enumeration of separate fragments of a body that is capable of processing the universal. Post-Borges, Kiš emphasizes how each of the cases he mentions is but a small part of the material indexed by his anonymous sages. A table in a local café can be the tiny key to the doors of the superimposed layers of a vast city. And one man can have the key that gives access to a world that ignores geo-political frontiers, that understands Europe as a unique cultural space beyond wars or the fall of empires. A cultural space that is always hospitable, because it only exists within the minds of those who walk and travel there. Unlike Borges, for whom history is unimportant, Zweig is keen to talk about how the First World War invented present-day frontiers. Mendel had lived his whole life in peace, without a single document to prove his original nationality or authorize his residency in the country where he was living. The news that war has broken out never reaches his bookish world and the postcards he continues to send to booksellers in Paris or London—those enemy capitals—suddenly attract the attention of the censor (a reader who is central to the history of the persecution of books, a reader who spends his time betraying readers). The secret police discover that Mendel is Russian and therefore a potential enemy. He loses his glasses in a skirmish. He is interned in a concentration camp for two years, during which time reading, his most beloved, pressing, perpetual activity, is interrupted. He is released thanks to important, influential clients, book collectors who know the man is a genius. But when he returns to the café, he finds he has lost the ability to read as he once did and thus spirals irreversibly towards eviction and death.

It is significant that he is a wandering Jew, a member of the People of the Book, that he comes from the East and meets misfortune and his end in the West, even though this only happens after dozens of years of unconscious assimilation, of being respected and even venerated by the chosen few who recognize that he is indeed exceptional. Mendel's relationship with printed information, Zweig tells us, caters to all his erotic needs. Like the ancient sages of Africa, he is a library man and his world is the non-material, accumulated energy that he shared.

The story is recounted by a sole survivor from the old days, when the café had a different owner and staff and stood for the world that disappeared between 1914 and 1918: an old woman who remembered Mendel. She is the memory of an existence sentenced to oblivion (if it weren't for the fact that a writer listens to her and records a testimony he later transforms into a story). Thanks to this process of evocation and research, to the critical distancing of time, the narrator, so like Zweig, echoes like an epiphany:

> In him, I had come close for the first time to the great mystery of the way what is special and overwhelming in our existence is achieved only by an inner concentration of powers, a sublime monomania akin to madness.

He is racked by shame. Because he has forgotten a model, a teacher. And a victim. The whole story builds up to that *recognition* and speaks obliquely of a great displacement: from the periphery in youth to a possible centre in maturity that has origins that should never have been forgotten. It is the story of a journey to those origins, a physical journey that encompasses another mnemonic and culminates in an homage. Generous and ironic, the narrator allows the illiterate old woman to keep the risqué volume that belonged to Mendel and that represents one of the few palpable traces of his passage through this world: "And I was the one," the story ends, "who ought to know that you create books solely to forge links with others even after your own death, thus defending yourself against the inexorable adversary of all life, transience and oblivion."

By paying homage to an itinerant bookseller from a world that has disappeared, Zweig acts like a historian as understood by Walter Benjamin: a collector, a rag-and-bone merchant. In this respect, Georges Didi-Huberman wrote in his essay, "Before Time": "A remnant not only provides a symptomatic support for ignorance—the truth about a repressed period of history, but also the very place and texture of the 'content of things,' of 'work on things.'" Funes' memory is like a garbage bin. The stories I have discussed, which might be taken to be about reading and memory, are in fact explorations of the relationship

between memory and forgetfulness. A relationship expressed through objects, volumes that are *containers*, the result of a kind of handicraft we call books that we read as remnants, as ruins of the texture of the past, of their ideas that survive. Because it is the fate of what is whole to be reduced to parts, to fragments, chaotic lists and examples that are still legible.

Books as objects, as things, bookshops as archaeological sites or junk shops or archives that resist revealing to us the knowledge they possess, that by their very nature refuse to occupy their rightful place in the history of culture, an often counter-spatial condition, opposed to any political organisation of space in terms of nations or states; the importance of inheritance, the erosion of the past, memory and books, non-material patrimony and its consolidation in materials that tend to decompose; Bookshop and Library as a two-faced Janus, or twin souls; police censorship, anti-frontier sites, bookshops as cafés and homes beyond the cardinal points of East and West, Orient and Occident; the lives and work of booksellers, whether sedentary or itinerant, isolated or members of a shared tradition, the tension between the unique and the serial; the power of encounters in a bookish context and their eroticism, latent sexiness; reading as obsession and madness but also as an unconscious drive or business, with its corresponding management problems and abuse of the workforce; numerous centres and infinite peripheries, the world as a bookshop and the bookshop as the world; irony and *gravitas*, the history of all books and specific books, with first names and surnames on jackets, of paper and pixels, bookshops at once universal and private: all that will be the subject of this book that until recently was in a bookshop or library or on a friend's bookshelf and now belongs, if only for a time, dear reader, to the library you call your own.

In other words, it has just left one *heterotopia* to enter another, with all the consequent changes of direction and differences in meaning. This book will *work* in this way: it will embrace the comfort of orderly reading and digressions and contradictions that disturb or threaten; it will re-create possible traditions and at the same time insist it only speaks of examples, exceptions from a map and a chronology of bookshops that it is impossible to re-create, that is made up of absence and oblivion, suggests analogies and synecdoche, a collection of glittering shards and leftover remnants from a future history or encyclopedia that can never be written.

I

Always a Journey

Every bookshop is a condensed version of the world. It is not a flight path, but rather the corridor between bookshelves that unites your country and its language with vast regions that speak other languages. It is not an international frontier you must cross but a footstep—a mere footstep—you must take to change topography, toponyms and time: a volume first published in 1976 sits next to one launched yesterday, which has just arrived; a monograph on prehistoric migrations cohabits with a study of the megalopolis in the twentieth-first century: the complete works of Camus precede those of Cervantes (it is in that unique, reduced space where the line by J.V. Foix rings truest: "The new excites and the old seduces"). It is not a main road, but rather a set of stairs, perhaps a threshold, maybe not even that: turn and it is what links one genre to another, a discipline or obsession to an often complementary opposite; Greek drama to great North American novels, microbiology to photography, Far Eastern history to bestsellers about the Far West, Hindu poetry to chronicles of the Indies, entomology to chaos theory.

You need no passport to gain entry to the cartography of a bookshop, to its representation of the world—of the many worlds we call *world*—that is so much like a map, that sphere of freedom where time slows down and tourism turns into another kind of reading. Nevertheless, in bookshops like Green Apple Books in San Francisco, in La Ballena Blanca (the White Whale) in the Venezuelan city of Mérida, in Robinson Crusoe 389 in

Istanbul, in La Lupa (the Magnifying Glass) in Montevideo, in L'Écume des Pages (the Foam of Pages) in Paris, in the Book Lounge in Cape Town, in Eterna Cadencia in Buenos Aires, in La Rafael Alberti in Madrid, in Casa Tomada (House Taken Over) in Bogotá, in Metales Pesados (Heavy Metals) in Santiago de Chile, in Dante & Descartes in Naples, in John Sandoe Books in London, in Literanta in Palma de Mallorca—in all these places I felt that I was stamping some kind of document, accumulating stamps that attested to my journey along an international highway, the most important or significant, the best or oldest or most interesting or simply the nearest bookshop when it suddenly started raining in Bratislava, when I needed a computer connected to the Internet in Amman, when I was finally forced to sit down and rest for a few minutes in Rio de Janeiro or when I wearied of so many shrines in Peru and Japan.

I picked up my first stamp in La Librería del Pensativo (the Thinker's Bookshop) in Guatemala City. I landed there at the end of July in 1998 when the country was still reeling from the outcry over Bishop Gerardi, who had been viciously murdered two days after he, the visible face of the bishopric's Human Rights Office, had launched the four volumes of the report "Guatemala: Never Again," which documented some 54,000 violations of basic human rights during almost thirty-six years of military dictatorship. They shattered his skull to the point that it was impossible to identify him by his facial features.

In those unstable months, when I switched abode four or five times, the cultural centre La Cúpula—comprising the gallery bar Los Girasoles (the Sunflowers), the bookshop and other shops—was most like home to me. La Librería del Pensativo sprang up in the nearby Antigua Guatemala in 1987 when the country was still at war, thanks to the tenacity of a feminist anthropologist, Ana María Cofiño, who had just returned from a long stay in Mexico. The familiar building on

the calle del Arco had once been a petrol station and car-repair workshop. Distant shots fired by the guerrillas, army or paramilitary echoed around the volcanoes surrounding the city. As happened and happens in so many other bookshops, and to a lesser or greater extent in bookshops throughout the world, the importing of titles hitherto unavailable in that central American country, support for national literature, launches, art exhibitions, and all that energy that soon linked the place to other newly inaugurated spaces, transformed El Pensativo into a centre of resistance. And of openness. After founding a publishing house for Guatemalan literature, they also inaugurated a branch in the capital that remained open for twelve years until 2006. I was happy there—although nobody there knows that.

After it closed, Maurice Echevarría wrote: "Now, with the presence of Sophos, or the gradual growth of Artemis Edinter, we have forgotten how El Pensativo sustained our lucidity and intellectual alertness after so many brains had been devastated."

I look for Sophos on the Internet: it is undoubtedly the place where I would spend my evenings if I still lived in Guatemala City. It is one of those spacious, well-lit bookshops with a restaurant and a family air that have proliferated everywhere: Ler Devagar in Lisbon, El Péndulo in Mexico City, McNally Jackson in New York, 10 Corso Como in Milan, or the London Review Bookshop in London, spaces that welcome communities of readers and soon transform themselves into meeting points. Artemis Edinter already existed in 1998, had done so for over thirty years, and now has eight branches; there must be a book in my library that I bought from one of them, but I do not remember which. In El Pensativo in La Cúpula I first saw the shock of hair, the face and hands of poet Humberto Ak'abal and learned by heart a poem he wrote about the ribbon the Mayans still use to tie up the bundles they carry on their heads that are sometimes three times their own weight and size ("For/us/Indians/the sky finishes/where *el mecapal* begins"); I watched a man crouch down to speak to his three-year-old son and saw the butt of a pistol sticking out from the belt of his jeans; I bought *Que me maten si . . .* by Rodrigo Rey Rosa, in a house edition, poor-quality paper I'd never touched before, which still reminds me of the paper my mother used to wrap my rolls in when I was a child, the feel of the thousand copies printed in the Ediciones Don Quijote printworks on December 28, 1996, almost a week after the democratic elections; I also bought there "Guatemala: Never Again," the single-volume précis of the original report's four volumes of death and hate: *the militarisation of children, multiple rapes, technology at the service of violence, psychosexual control of soldiers,* all that is contrary to what a bookshop stands for.

I found I had a *mappa mundi* rather than a passport the day I finally spread out all those stamps on my desk (visiting cards, postcards, notes, photographs, prints I had been putting in folders after each trip, anticipating the moment when I would begin this book). Or rather a map of my world. And consequently subject to my own life: how many of those bookshops must have closed their doors, changed address, multiplied, or must now be transnational or have reduced staff or opened a .com domain. A necessarily incomplete map criss-crossed by the length of my journeys, where huge areas remained unvisited and undocumented, where tens, hundreds of significant bookshops had yet to be noted (collected), though it nevertheless represents a possible overview of an ever-changing twilight scenario, of a phenomenon that was crying out to be analysed, written up as history, even if it would only be read by others who have also sat in bookshops here and thereabouts, so many embassies without a flag, time machines, *caravanserai*, pages of a document no state can ever issue. Because bookshops like El Pensativo have disappeared or are disappearing or have become a tourist attraction in countries across the world, have opened a website or been subsumed into a bookshop chain with the same name and then are inevitably transformed, adapting to the volatile—and intriguing—signs of the times. And here, before me, lay a collage that evoked what Didi-Huberman has described in *Atlas. How to carry the world on your back,* where—just as in the passageways of a bookshop—"the *affective* as much as the *cognitive* element" has equal value on my desktop between "*classification* and *disorder*" or, if you prefer, between "reason and imagination" because "tables act as operational fields to *disassociate*, disband, destroy," and to "*agglutinate*, accumulate and set out" and, consequently, "they gather together heterogeneities, give shape to multiple relationships": "where heterogeneous times and spaces continually meet, clash, cross or fuse."

The history of bookshops is completely unlike the history of libraries. The former lack continuity and institutional support. As private entrepreneurial responses to a public need they enjoy a degree of freedom, but by the same token they are not studied, rarely appear in tourist guides and are never the subject of doctoral theses until time deals them a final blow and they enter the realm of myth. Myths like St Paul's Churchyard, where—as I read in Anne Scott's *18 Bookshops*—the Parrot was one of thirty bookshops and its owner, William Apsley, was not only a bookseller but also one of Shakespeare's publishers, or the rue de l'Odéon in Paris, which nurtured Adrienne Monnier's La Maison des Amis des Livres and Sylvia Beach's Shakespeare and Company. Myths like Charing Cross Road, the intergalactic avenue, London's bibliophile street *par excellence*, immortalized in the best non-fiction book I have read on bookshops, *84, Charing Cross Road* by Helene Hanff (where, as in any shop selling books, bibliophile passion is shot through with human feelings, and drama coexists with comedy), a first edition of which I was excited to see on sale for £250 in the window of Goldsboro Books, an establishment that specializes in selling signed first editions, very close to the same

Charing Cross Road where nobody could tell me where I might find Hanff's bookshop. Myths like the bookshop *dei Marini*, later the Casella, that was founded in Naples in 1825 by Gennaro Casella and then inherited by his son Francesco, who at the turn of the twentieth century invited to his premises people like Filippo T. Marinetti, Eduardo de Filippo, Paul Valéry, Luigi Einaudi, George Bernard Shaw or Anatole France, who stayed in the Hotel Hassler del Chiatamone, but treated the bookshop as his front room. Myths like Moscow's Writers' Bookshop that in the 1900s and the early 1920s made the most of a brief interlude of revolutionary freedom and gave readers a centre of culture managed by intellectuals. The history of libraries can be told in minute detail, ordered by cities, regions and nations, respecting the frontiers that are sealed by international treaties and drawing on specialized bibliographies and individual library archives that fully document the development of stocks and cataloguing techniques and house minute-books, contracts, press cuttings, acquisition lists and other papers, the raw material for a chronicle backed by statistics, reports and timelines. The history of bookshops, on the other hand, can only be written after recourse to photograph and postcard albums, a situationist mapping, short-lived links between shops that have vanished and those that still exist, together with a range of literary fragments and essays.

When I was sorting out my visiting cards, leaflets, triptychs, postcards, catalogues, snapshots, notes and photocopies, I came across several bookshops that did not fit any geographical or chronological criteria, could not be explained in terms of the stopovers and paths I was tracing for others, however conceptual and transversal these might be. I am referring to bookshops that specialize in travel, a paradox in itself, because every bookshop is an invitation to travel, and itself represents a journey. But the latter *are* different. The word "specialize" points to their peculiarity. Like children's bookshops, comic shops, antiquarian bookshops and those trading in rare books. Their specialist focus is evident in the way they categorize their books: not by genre, language or academic discipline, but by geographical area. This principle is taken to an extreme in Altaïr, whose main shop in Barcelona is one of the most absorbing bookish spaces I know. There they group poetry, fiction and essays according to country and continent, so you find them next to the relevant guidebooks and maps. Travel bookshops are the only ones where cartography outshines prose and poetry. If you follow the itinerary suggested by Altaïr, you pass by the window display to a noticeboard of messages posted by travellers. Behind that sits a collection of the shop's house magazines. Then come novels, histories and themed guidebooks on the subject of Barcelona in a pattern followed by most of the world's bookshops, as if the logic of necessity meant one must move from the immediate and local to what is most remote: the universe. Consequently, the world is next, also arranged according to criteria of distance: from Catalonia, Spain and Europe to the remaining continents, the world spreads across the two floors of the shop. Maps of the world are downstairs and beyond them, at the back, a travel agency. Noticeboard, magazines and all that reading matter can lead to only one outcome: setting out.

Ulyssus in Girona carries the secondary name of Travel Bookshop, and like the founders of Altaïr, Albert Padrol and Josep Bernadas, its owner, Josep María Iglesias, sees himself first and foremost as a traveller and secondly as a bookseller or publisher. Ulysses, the Paris bookshop, likewise has Catherine Domain at the helm. A writer and explorer, Domain obliges her staff to travel with her every summer to the casino in Hendaye. By symbolic extension, this kind of establishment is usually full of maps and globes of the earth: in Pied à Terre in Amsterdam, for example, there are dozens of globes that observe you on the sly as you hunt for guides and other reading matter. Its slogan could not be more insistent: "The traveller's paradise." Deviaje (Travelling), the Madrid bookshop, emphasizes its character as an agency: "Bespoke travel, bookshop, travel accessories." The ordering does not alter the end product, because the truth is that travel bookshops throughout the world are also stores that sell practical travel items. Another Madrid shop, Desnivel (Uneven), specializing in exploration and mountaineering, sells GPS trackers and compasses. The same is true of Chatwins in Berlin, which devotes a good part of its display space to Moleskine notebooks, the mass-produced reincarnation of the artisan-made jotters that Bruce Chatwin used to buy in a Paris shop until the family in Tours that manufactured them stopped doing so in 1986. He relates this in *The Songlines*, a book published the following year.

Chatwin's funeral was held in a West London church, though in 1989 his ashes were scattered by the side of a Byzantine chapel in Kardamyli, one of the seven cities Agamemnon offers Achilles in return for the renewal of his offensive against Troy in the southern Peloponnese, and near the home of one of his mentors, Patrick Leigh Fermor, a travel writer and, like him, member of the Restless Tradition. Thirty years earlier, a young man from the provinces by the name of Bruce Chatwin, without trade or income, had arrived in London to work as an apprentice at Sotheby's, unaware of his future as a travel writer, a mythomaniac and, above all, a myth in himself. He was unaware, too, that he would give his name to a bookshop in Berlin. Two bookshops stand out among the many Chatwin might have discovered when he arrived in the city at the end of the 1950s: Foyles and Stanfords. One generalist and the other specializing in travel. One full of books and the other awash with maps.

In the middle of Charing Cross Road, Foyles' fifty kilometres of shelves make up the world's greatest print labyrinth. In that period it became a tourist attraction because of its size and the absurd ideas put into practice by its owner, Christina Foyle, who turned the place into a monstrous anachronism in the second half of the twentieth century. Ideas like refusing to use calculators, cash registers, telephones or any other technological advances to process sales and orders, or arranging books by publishing house and not by author or genre, or forcing her customers to stand in three separate queues to pay for their purchases, or sacking her employees for no good reason. Her chaotic management of Foyles—which was founded in 1903— lasted from 1945 to 1999. Her eccentricities can be explained genetically: William Foyle, her father, committed his very own lunacies before handing the shop over to his daughter. Conversely, Christina must be credited for the finest initiative taken by the bookshop in all its history: its renowned literary lunches. From October 21, 1930 to this day half a million readers have dined with more than a thousand authors, including T. S. Eliot, H. G. Wells, George Bernard Shaw, Winston Churchill and John Lennon.

Such Notoriety now belongs to the past (and to books like this): in 2014 Foyles was transformed into a large modern bookshop and moved to the adjacent building at 107 Charing Cross Road. The reshaping of the old Central Saint Martin's

College of Art and Design was the responsibility of the architects of Lifschutz Davidson Sandilands, who met the challenge of designing the largest bookshop built in Britain in the twenty-first century. They created a large, empty central courtyard suffused with bright white light reinforced by huge lamps that punctuate the vast, diaphanous text, which is surrounded by stairs that go up and down like so many subordinate clauses. A cafeteria—which is always buzzing—is at the top, next to an exhibition room equipped for trans-media projects and the main presentation room. When you walk in you are greeted by a sign at ground level: "Welcome, book lover, you are among friends." What would Christina see if she raised her head? She would see an entire wall commemorating her crowded lunches.

"Explore, describe, inspire" is Stanfords' slogan, as I am reminded by the bookmark I keep as a souvenir of one of my visits to that shop. Although the business was founded in that same Charing Cross Road where Foyles still survives, its famous Covent Garden headquarters in Long Acre opened its doors to the public in 1901. By then, Stanfords had already forged a strong link with the Royal Geographical Society by virtue of producing the best maps in an era when the expansion of British

colonialism and an increase in tourism had led to a massive rise in the printing of maps. Although you can also find guidebooks, travel literature and related items on the store's three levels, whose floors are covered by a huge map (London, the Himalayas, the World), cartography plays the lead role. Even the bellicose variety: from the 1950s to the 1980s the basement was home to the maritime and military topography department. I remember I visited Stanfords because someone told me, or I read somewhere, that Chatwin bought his maps there, though there is no record of him ever having done so. The shop's list of distinguished customers comprises everyone from Dr Livingstone and Captain Robert Scott to Bill Bryson or Sir Ranulph Fiennes, one of the last living explorers, to say nothing of Florence Nightingale, Cecil Rhodes, Wilfred Thesiger or Sherlock Holmes, who ordered the map of the mysterious moor that enabled him to solve his case in *The Hound of the Baskervilles* from Stanfords.

Foyles has five branches in London and one in Bristol. Stanfords has shops in Bristol and Manchester, as well as a small space in the Royal Geographical Society that only opens for events. Chatwin missed by a couple of years the opportunity to experience Daunt Books, a bookshop for travelling readers, whose first shop—an Edwardian building on Marylebone High Street naturally lit by huge plate-glass windows—opened in 1991. The store was a personal project of James Daunt, the son of diplomats, and thus used to moving house. After a stay in New York, Daunt decided he wanted to dedicate himself to his two passions in life: travel and books. Daunt Books is now a London chain with six branches. Au Vieux Campeur has sold maps, and travel books and guides as well as hiking, camping and climbing equipment from 1941 and now boasts a grand total of thirty-four establishments across France. Such is the way of Moleskine logic.

At the end of the nineteenth century and at the beginning of the twentieth, many amateur and professional artists took up the habit of travelling with sketchbooks that had thick enough paper to cope with watercolours or India ink and sturdy covers to protect drawings and paintings from the elements. They were manufactured in different parts of France and sold in Paris. We now know that Wilde, Van Gogh, Matisse, Hemingway and Picasso used them, but how many thousands of anonymous travellers did also? Where might their *Moleskines* be? Chatwin gives them that name in the Australian book we mentioned and it was what encouraged Nodo & Nodo, a small Milanese firm, to launch five thousand copies of these Moleskine notebooks onto the market in 1999. I remember experiencing, on seeing some of them or the limited editions following on from that first printing in a Feltrinelli bookshop in Florence, an immediate surge of fetishist pleasure, the kind that *recognition* brings. It is what any committed reader feels on walking into Lello in Oporto or City Lights in San Francisco. For years you were forced to travel to buy a Moleskine. It was not necessary to go to a Paris bookshop, though that did not mean you could find them in any bookshop in the world. In 2008 they were supplied to some 15,000 shops in over fifty countries. To cope with demand, production was moved to China, although the design remained Italian. Before 2009 I had to go to Lisbon if I wanted to visit Livraria Bertrand, the oldest bookshop in the world, which fleetingly opened a branch in Barcelona, the city where I live, and serial commercial expansion won yet another victory—its *nth*—over that old idea, now almost without a body to flesh it out: atmosphere.

II
Athens: A Possible Beginning

One can walk around and read Athens as if it were a strange souk of bookshops. Of course, the strangeness is a consequence of the decadent atmosphere and palpable feeling of antiquity rather than of the language in which shop names and shelf labels are written, not to mention book titles and author names. For a Western reader, the East begins where unknown alphabets start to be used: Sarajevo, Belgrade and Athens. On the shelves of bookshops in Granada and Venice no trace remains of the alphabets of anyone who arrived in those cities from the East in the remote past: we read all that translated into our languages and have forgotten that theirs were also translations. The centrality of ancient Greek culture, philosophy and literature can only be understood if one considers its position astride the Mediterranean and Asia, between the Etruscans and Persians, opposite the Libyans, Egyptians or Phoenicians. Its situation as an archipelago of embassies. Or radial aqueduct. Or network of tunnels between different alphabets.

After a long search on the Internet, inspired by the card of one of the establishments I have kept from 2006, I finally find a reference in English to what I am looking for: Books Arcade, Book Gallery or Book Passage, a succession of twenty spaces with wrought-iron gates that are home to forty-five publishers, including Kedros and Publications of the National Bank. I made notes on the ways bookshops relate to libraries sitting on one of the many armchairs in those passageways, under a ceiling fan that sliced through the heat in slow motion. Because the

Pesmazoglou Arcade—another of its names, and a reference to one of the streets giving access—is located opposite the National Library of Greece.

The Tunnel opposite the Building. The Gallery with no inaugural date opposite the Monument recorded in minute historical detail: neo-classical in style, financed from the diaspora by the Vallianos brothers. The first stone of the National Library was laid in 1888 and it was inaugurated in 1903. It conserves and houses some 4,500 ancient Greek manuscripts, Christian codices and important documents from the Greek revolution (it was not for nothing that the idea behind it apparently came from Johann Jakob Mayer, a lover of Hellenic culture and comrade-in-arms of Lord Byron). But any library is more than a building: it is a bibliographical collection. The National Library was previously lodged in the orphanage in Aegina, the baths in the Roman Market, the church of St Eleftherios and the University of Otto; over the next few years it will transfer to a grandiose new building on the seafront, designed by architect Renzo Piano. The present Library of Alexandria is a far cry from the original: although its architecture is spectacular, although it converses with the nearby sea and 120 alphabets are inscribed on its reflective surface, although tourists will come from all over the world to gaze at it, its walls do not yet contain sufficient volumes for it to be considered the reincarnation of the building that lends it its mythical name.

The shadow of the Library of Alexandria is so powerful it has eclipsed every other previous, contemporary and future library, and has erased from collective memory the bookshops that nourished it. Because it was not born in a void: it was the main customer of book traders in the eastern Mediterranean in the third century BC. The Library cannot exist without the bookshop that has in turn been linked from the outset to the publishing house. The book trade had already developed before the fifth century BC; by this date—when the written was beginning to prevail over the oral in Hellenic culture—the works of the main philosophers, historians and poets we today think of as the classics were known in a large part of the eastern Mediterranean. Athenaeus quotes a lost work by Alexis, from the fourth century BC, entitled *Linos*, where the hero says to young Hercules:

> "Take one of these beautiful books. Look at the titles in case one is of interest. Here you have Orpheus, Hesiod, Keralis, Homer and Epicharmus. There you'll find plays

and everything you might want. Your choice will reveal your interests and taste."

In the event, Hercules chooses a cookery book and does not meet his companion's expectations. Because the book trade includes every kind of text and reading taste: speeches, poems, jottings, technical or law books, collections of jokes. And it also encompasses every level of quality: the first publishing houses comprised groups of copyists on whose ability to concentrate, to be disciplined and rigorous and on whose degree of exploitation depended the number of changes and mistakes in the copies that would eventually be put into circulation. To optimize time, someone dictated and the rest was transcribed, thus allowing Roman publishers to launch onto the market several hundred copies simultaneously. In his exile, Ovid consoled himself with the thought that he was "the most read writer in the world" since copies of his works reached the furthest boundary of the empire.

In his *Libros y libreros en la Antigüedad* (a shortened version of H.L. Pinner's *The World of Books in Classical Antiquity* that was only published after his death), Alfonso Reyes talks of "book traders" when he refers to the first publishers, distributors and booksellers, like Atticus, Cicero's friend, who was involved in every facet of the business. Apparently the first Greek and Roman bookshops were either itinerant stalls or huts where books were sold or rented out (a kind of mobile library) or spaces adjacent to the publishers. "In Rome bookshops were well known, at least in the days of Cicero and Catullus," writes Reyes. "They were located in the best commercial districts, and acted as meeting places for scholars and bibliophiles." The Sosii brothers, publishers of Horace, Secundus, one of the publishers of Martial and Atrectus, among many other entrepreneurs, managed premises in the vicinity of the Forum. Lists were posted on the door advertising the latest books. And for a small amount one could consult the most valuable volumes, in a kind of fleeting loan. The same happened in big cities in the empire, like Rheims or Lyons, whose excellent bookshops surprised Pliny the Younger when he saw that they too sold his books.

The sale and purchase of beautiful copies increased, as did the acquisition of volumes by weight so that wealthy Romans could cover a wall with a pretence to culture and boast about their libraries. Private collections, often in the hands of bibliophiles, were directly fed by bookshops and were a model for public collections, namely libraries, which sprang up in tyrannies, not democracies: the first two are attributed to Polycrates, the Tyrant of Samos, and Pisistratus, the Tyrant of Athens. Libraries are power: in 39 BC, Gaius Asinius Polio founded the Library of Rome with booty from his campaign in Dalmatia. Greek and Roman titles were exhibited there for the first time publicly and together. Four centuries later there were twenty-eight libraries in the capital of the later Roman Empire. Now they are ruins like the library in Pergamum or the Palatine Library.

The Library of Alexandria was seemingly inspired by Aristotle's private library and was probably the first in history to have a cataloguing system. The dialogue between private and public collections, between the Bookshop and the Library, is therefore as old as civilisation itself, but the balance of history always inclines towards the latter. The Bookshop is light; the Library is heavy. The levity of the present continuous is counterpoised by the weight of tradition. Nothing could be more alien to the idea of a bookshop than heritage. While the Librarian accumulates, hoards, at most lends goods out for a short while—which thereby cease to be such or have their value frozen—the Bookseller acquires in order to free himself from what he has acquired; he sells and buys, puts into circulation. His business is *traffic* and *transit*. The Library is always one step behind: looking

towards the past. The Bookshop, on the other hand, is attached to the sinews of the present, suffers with it, but is also driven by an addiction to change. If history ensures the continuity of the Library, the future constantly threatens the existence of the Bookshop. The Library is solid, grandiose, is tied to the powers-that-be, to local authorities, to states and their armies: as well as despoiling the patrimony of Egypt, "Napoleon's army carried off around 1,500 manuscripts from the Austrian Lowlands and another 1,500 mainly from Bologna and the Vatican," writes Peter Burke in his *A Social History of Knowledge*, in order to feed the voracious libraries of France. Conversely, the Bookshop is liquid, provisional, lasts only as long as its ability to sustain an idea over time with minimal changes. The Library is stability. The Bookshop distributes; the Library preserves.

The Bookshop is in perpetual crisis, subject to the conflict between *novelty* and *stocks*, and precisely for that reason finds itself at the centre of the debate over cultural canons. Great Roman authors were aware that their influence depended on the public's access to their intellectual production. The figure of

Homer is located in the two centuries prior to the consolidation of the bookselling business and his centrality to the Western canon is directly related to the fact that he is one of the Greek writers of whose work we have preserved the most fragments. That is, he is one of the most copied. He is also one of the most disseminated, sold, gifted, stolen and purchased by collectors, general readers, booksellers, bibliophiles and library administrators. Our concept of cultural tradition, our list of writers and key titles depends on papyrus and parchments, rolls and codices from Greek and Roman bookshops, on the textual capital they put into circulation, provisionally confined in public and private spaces, the majority of which was destroyed in countless wars and fires and moves. The location of bookshops is fundamental to the structuring of these canons: there was a time when Athens and Rome were the centre of all possible worlds. We have built our entire subsequent culture on these lost, indemonstrable capitals.

SCRIPTORIUM MONK AT WORK. (From *Lacroix*.)

The traffic in books shrank after the fall of the Roman Empire. Medieval monasteries continued the task of spreading written culture, through copyists, at the same time as, thanks to Islam, paper ended its long journey from China, where it was invented, to the south of Europe. Parchment was so expensive that texts were often erased so others could be put in their place: there are few metaphors as powerful as that of the palimpsest to represent the way culture is transmitted. In the Middle Ages, a book could have some hundred handwritten copies, be read by several thousands and listened to by many more since orality was once again more important than individual reading. That does not mean that the bookselling trade came to a halt. The noble and ecclesiastical classes needed, after all, to read, as did the students who relied on printed texts in increasing numbers, the oldest European universities (Bologna, Oxford, Paris, Cambridge, Salamanca, Naples . . .) having been founded between the eleventh and thirteenth centuries. As Alberto Manguel has written in *A History of Reading*:

> From roughly the end of the twelfth century, books became recognized as items of trade, and in Europe the commercial value of books was sufficiently established for money-lenders to accept them as collateral; notes recording such pledges are found in numerous medieval books, especially those belonging to students.

The pawning of books was a constant from that moment on, until Xerox popularized photocopying in the middle of the last century. Photocopying shops coexist in the neighbourhood of the National Library of Greece and adjacent Academy of Athens, with universities, publishing houses, cultural centres and the most compact part of the souk of bookshops, because all these institutions feed on each other. I remember reading an edition of poetry by Cavafy that I'd been carrying in my ruck-sack in the spacious piano bar in the Ianos bookshop, a link in a *chain of civilization*, with its mahogany-coloured shelves and white-on-apple-green labels, simply because I could not under-stand a single one of the volumes around me. I also remember spending hours between the dark wooden shelves in the Politeia bookshop; browsing, among the thousands of books in Greek, through the few hundred that had been published in English. Divided between two floors and a basement, the premises have four doors of entry. It is one of those over-lit spaces: countless rectangles of light, from barely six spotlights, make the covers, titles and floor gleam. *Politeia* means "community of citizens."

Finally, I walked into the Librairie Kauffmann. Not only because it is *the* French bookshop in Athens and hence stocks books that I can read, but also because it is one of those bookshops to get a stamp on your imaginary bookshop passport. The black-and-white image of the shop's launch is striking: dated 1919, it shows a kiosk attended by a woman dressed in traditional fashion, her head partially covered, above her the words: "Librairie Kauffmann." Hermann Kauffmann began his business with a street stall selling second-hand French books. Ten years later, he set up in premises on Zoodochos Pigis Street, which eventually grew into a kind of large apartment with views over the avenue, and incorporated new titles into his shelves thanks to an agreement with the publisher Hachette. Soon it became the place to visit for the most enlightened people in Athens, who came to stock up on French reading-matter whilst their children bought textbooks and course reading for their French-speaking schools and academies. A diploma granted to Kauffmann in 1937 by L'Exposition Internationale des Arts et Techniques dans la Vie Moderne in Paris hangs on the staircase wall next to photographs of Frida Kahlo and André Malraux. With Hachette's help, he created the Hellenic Distribution Agency. After his death in 1965, his widow took over the firm and was behind important initiatives, like the "Confluences" collection of Greek literature translated into French, or the publication of the *Dictionnaire français-grec moderne*—something that should always exist in a bookshop specializing in a foreign language: a dictionary that is at least bilingual.

The Kauffmann web page no longer works. There is no sign on the web that the bookshop is still open. After several futile searches, I salvage the orange card from that trip, with a tree embossed above Greek and Latin characters, like a murky archipelago at the bottom of the sea. And I dial the number. Two or three times. Nobody picks up the telephone. As I wander from one search engine to the next I finally find photographs I did not want to see. One shows the Pesmazoglou Arcade—or Book Gallery—burnt to the ground during the riots at the beginning of 2012 because it was home to private enterprises, including a branch of the National Bank's publishing house. On the other hand, although the international press initially reported that the library had also been burned down, it was not attacked or ravaged by fire: public and ancient, with an inaugural date and plans to move. Its past and future guaranteed as much as anything can be, it remains standing.

III
The Oldest Bookshops in the World

As well as being old, a bookshop must look the part. When you go into the Livraria Bertrand, at 73 rua Garrett in Lisbon, very close to the Café Brasileira and its Fernando Pessoa statue in the heart of the Chiado, the B on the red background of the logo proudly displays a date: 1732. Everything in the first room points to the venerable past that this date highlights: the display cabinet of extraordinary books; the extending steps or wooden ladders that give access to the highest shelves of some of the ancient bookcases; the rusty plaque that dubs the place where you stand "Sala Aquilino Ribeiro," in homage to one of its most distinguished customers, who was a regular like Oliveira Martins, Eça de Queirós, Antero de Quental or José Cardoso Pires; and above all the certificate from the Guinness World Records that attests to it being the oldest functioning bookshop in the world.

A functioning, moreover, that has been uninterrupted and is well documented. Books have been sold intermittently since 1581 at 1 Trinity Street, Cambridge, to famous customers like William Thackeray and Charles Kingsley, but for long stretches the premises were exclusively the seat of Cambridge University Press, with no direct sales to the public. Conversely, caught in a quagmire of the absence of reliable documents, we find Matras in the city of Kraków—still called Gebether and Wolff by elderly locals—the mythical origins of which go back to the seventeenth century (when the book dealer Franz Jacob Mertzenich opened a bookshop in 1610 that did not close until 1872), and which was the site of a renowned literary salon at

the turn of this century and now hosts important literary events within the framework of a UNESCO city of literature. That is why the Librairie Delamain in Paris, which opened its doors in the Comédie Française in 1700 or 1701—according to sources—and did not move to rue Saint-Honoré until 1906, may genuinely be the world's oldest bookshop, though I imagine it cannot prove such a felicitous uninterrupted period of bookselling. Winchester's P&G Wells bookshop does appear to be the oldest bookshop in the United Kingdom and quite possibly the oldest in the world with single premises that are emphatically independent (it opened a branch at the end of the twentieth century at the University of Winchester). Receipts for the purchase of books from 1729 have been preserved and constant bookselling activity in the shop on College Street appears to date back to the 1750s. In 1768, Hodges Figgis began to deal in books. The oldest bookshop in Ireland and still active, it is also the largest with a stock of 60,000 books. It is equally the most Dublinesque of bookshops, because it appears in that most Dublinesque of books, which is not Joyce's *Dubliners* but *Ulysses*, ("She, she, she. What she? The virgin at Hodges Figgis' window on Monday looking in for one of the alphabet books you were going to write."). Hatchards, which opened its doors in 1797 and has never shut them since, is London's oldest bookshop. Its aristocratic building at 187 Piccadilly, with the portrait in oils of its founder, John Hatchard, gives the establishment a suitably antique patina. It now belongs to the Waterstones chain, but it hasn't lost any of its plush-carpet identity: unlike more commercial bookshops, it still sells novels on the first floor and reserves the ground floor for hardback history books, biographies and current affairs, which are still purchased by customers on their way to the Royal Academy or their Jermyn Street tailors. In recent years the shop has initiated a subscription service that, in our era of algorithms, employs three expert readers to study the tastes of subscribers and dispatch a selection of books to them on a regular basis. Mary Kennedy, who was my guide to the shop's history and hidden corners, told me proudly, "They all have the right to return the titles they don't like but we have only ever had one return."

The only really important nineteenth-century bookshop I have visited is perhaps the Librería de Ávila—opposite the church of San Ignacio and very close to the Colegio Nacional de Buenos Aires. It was supposedly founded in 1785, the year when a shop was established on that same corner selling food, alcohol and books. If P&G Wells printed books for Winchester

College, its contemporaneous bookshop in Buenos Aires was linked to the nearby educational institution even by name: Librería del Colegio. There are no extant documents relating to the Librería del Colegio at the same address until 1830, when it is frequently mentioned in the press, since it hosted conversations led by Sarmiento with Estrada, Hernández, Alberdi, Aristóbulo del Valle, Groussac, Avellaneda, Perito Moreno and other intellectuals who now have streets named after them. The present building on this corner was not built until 1926, where thirteen years later the Editorial Sudamericana was founded, existing side by side with the Librería del Colegio until 1989. It was shut for four years before Miguel Ángel Ávila bought the business and renamed it after himself. The names of these shops changed much less than those of the two streets where they shared a corner: San Carlos and Santísima Trinidad, Potosí and Colegio, Adolfo Alsina and Bolívar, at the last count.

It says "*Antiguos Libros Modernos*" on the façade. On my first visit to Buenos Aires, in July 2002, I bought a few copies of the magazine *Sur* in its basement. Touching old books is one of the few tactile experiences that can connect you to a distant past. Although the concept of the antiquarian bookshop belongs to the eighteenth century as a result of the corresponding growth of disciplines like history and archaeology, in the previous two centuries it had been developed by bookbinders and sellers who worked as much with printed books as handwritten copies. The same can be said of the catalogues of printers and publishers that evolved, from simple lists of publications, into small but sophisticated books. I have never so much as touched one of these relics. Or even a book that was not printed.

Svend Dahl in his *A History of the Book* states that manuscripts prevailed over printed books, in the first years of printing, by virtue of a veneer of prestige, as was the case with papyrus over parchment, or in the 1960s, with handmade books over ones that were machine-set. In the beginning, the printer was the publisher: "But itinerant sellers soon appeared who went from city to city offering books they had bought from printers." They hawked the titles they were carrying on the streets and in the taverns where they stayed and set up a nomadic market. Some also had permanent stalls in big cities. From the sixteenth century, copies of the same book could be bought in the thousands. And their readers were in the hundreds of thousands: more than a hundred thousand different printed books began to flood Europe over the next hundred years. A double system of classifying and displaying books was developed using filing boxes and cards or bookcases, it being usual for the books to be unbound so customers could choose the kind of binding they wanted for their individual copy. Hence those whimsical collections of books that only have the covers their owners selected in common. Some can be found intact in the basement of the Librería de Ávila and in the second-hand bookshops on the Avenida de Mayo.

MESSRS LACKINGTON ALLEN & C.º
TEMPLE OF THE MUSES, FINSBURY SQUARE.

What were the bookshops of the eighteenth century like, when Bertrand Livreiros, Hatchards and the Librería del

Colegio opened their doors in Lisbon, London and Buenos Aires, respectively? As one can see from the seventeenth- and eighteenth-century engravings studied by Henry Petroski in *The Book on the Bookshelf*, a detailed history of how we arrange our books and the evolution of the bookshelf, the bookseller sat behind a large desk to manage his business, which was often physically linked to the printers or publishers on which it depended, and was surrounded by a display from a large archive of sewn, though not bound, folders that were the actual book-shops. The boxes were often part of the counter, as one can see in a famous engraving of the Temple of the Muses, perhaps the most legendary and beautiful eighteenth-century bookshop, located in Finsbury Square and run by James Lackington, who, refusing to destroy unsold books, sold them off cheaply in accordance with what he understood as his professional mission. He wrote: "Books are the key to knowledge, reason and happiness, and everyone, no matter what their economic background, social class or sex, has the right to have access to books at cheap prices."

Goethe's is one of the finest written testimonies to book-shops; on September 26, 1786, he jotted in his *Italian Journey*:

At last I have acquired the works of Palladio, not the original edition with woodcuts, but a facsimile with copperplate engravings published by Smith, an excellent man who was formerly English consul in Venice. One must give the Eng-lish credit for having so long appreciated what is good and for their munificence and remarkable skill in publicizing it.

On the occasion of this purchase, I had entered a book-shop which, in Italy, is a peculiar place. The books are all in stitched covers and at any time of day you can find good company in the shop. Everyone who is in any way connected with literature—secular clergy, nobility, artists—drop in.

You ask for a book, browse in it or take part in a conversation as the occasion arises. There were about half a dozen people there when I entered, and when I asked for the work of Palladio, they all focused their attention on me. While the proprietor was looking for the book, they spoke highly of it and gave me all kinds of information about the original edition and the reprint. They were well acquainted with the work and with the merits of the author. Taking me for an architect, they complimented me on my desire to study this master who had more useful and practical suggestions to offer than Vitruvius, since he had made a thorough study of classical antiquity and tried to adapt his knowledge to the needs of our times. I had a long conversation with these friendly men and learned much about the sights of interest in the town.

The first sentences tell of the fulfilment of a wish: the aim of every visit to a bookshop. The final sentences, the acquisition of knowledge that is not to be found in the books themselves, but in the people in their vicinity. What most surprises the erudite German traveller is the fact that all the books are bound and accessible, so visitors can establish dialogues as much with the books as among themselves. Binding didn't become standard in Europe until the requisite machines began to function around 1823. When bookshops began to look like libraries, because they offered finished products and not half-made books, what surprised Goethe was the handmade bindings. In *A Sentimental Journey Through France and Italy* (1768), Laurence Sterne enters a bookshop on the Conti quayside to buy "a collection of Shakespeare," but the bookseller tells him he does not have one. The traveller indignantly picks up a copy on the table and asks, "And what about this?" And the bookseller explains that it is not his, that it belongs to a count, who has sent it to be bound: he is an "*esprit fort*," he explains, "fond of English books" and hob-nobbing with islanders.

When Chateaubriand went to Avignon in 1802 after being tipped off about pirated copies of four volumes of his *The Genius of Christianity*, he recounts in his memoirs that "By going from bookshop to bookshop, I unearthed the counterfeiter, to whom I was unknown." Every city had many and we have preserved no record of most. We tend to think of literature as an abstraction when the truth is that it is an infinite network of objects, bodies, materials and spaces. Eyes that read, hands that write and turn pages and hold tomes, cerebral synapses, feet that seek out bookshops and libraries, or vice versa, biochemical desire, money to purchase, paper and cardboard, stocked shelves, pulped timber and vanished forests, more eyes and hands that drive lorries, load boxes, order volumes, browse, peer and leaf, contracts, letters, numbers and photographs, warehouses, prem-ises, square metres of cities, characters, screens, wonders in ink and pixels.

The word *poiein* that in ancient Greek meant "making" is the linguistic root of "poetry." In *The Craftsman*, the sociologist Richard Sennett has explored the intimate relationship between hand and eye: "Every good craftsman conducts a dialogue between concrete practices and thinking; this dialogue evolves into sustaining habits, and these habits establish a rhythm between problem solving and problem finding." He focuses especially on carpenters, musicians, cooks, luthiers, people we generally understand to be *craftsmen*, but the truth is his reflections can be extended both to the endless craftsmen who have always collaborated in the making of books (paper-makers, typographers, printers, binders, illustrators) and the actual bodies of readers, their dilating pupils, ability to concentrate, bodily posture, digital memory (in their fingertips). Writing itself, inasmuch as it is calligraphy—that is, manufacture—is even subject to a discipline of perfection in civilizations like the Arab and Chinese. And the move from writing by hand to keying in is still very recent in the history of culture. Although he does not intervene directly in the creation of the object, the bookseller can be understood as the *craftsman reader*, that person who, after the 10,000 hours that, according to various studies are necessary to become expert in a practical skill, is able to combine work with excellence, manufacture with poetry.

Romano Montroni, who for decades worked in Feltrinelli's in the Piazza di Porta Ravegnana, Bologna, writes in "The Bookseller's Decalogue" that "the customer is the most important person in the enterprise," and places dusting at the centre of activity in a bookshop: "One must dust every day and everyone must do it!" he exclaims in *Vender el alma: El oficio de librero.* "Dust is a vitally important issue for a bookseller. He dusts up and down and clockwise in the first half-hour every morning. While doing so, the bookseller memorizes where the books are and gets to know them *physically.*"

Some of the world's bookshops carefully nurture their tactile dimension, so that paper and wood bear witness to a tradition of craftsman readers. In England, for example, the three branches of Topping and Company were furnished with shelving made by local carpenters, and the small signs labelling the sections and the cards recommending titles are handwritten. The Bath branch's well-stocked poetry section shows how important it is for a bookshop to cherish and develop the interests of the local community. "People here have a great fondness for poetry," Saber Khan, one of its staff, told me, "and we stock the largest selection of poetry in the country." Readers, like carpenters, are different in each locality: the branches of Topping and Company "have their own identity, like brothers and sisters, but in every one coffee is free, because you can't deny anyone their cup of coffee." I saw readers who sat for hours on wooden chairs at wooden tables. There was a bed and bowl for the dog who roamed around the shop, his home and ours. Its slogan, "A proper old-fashioned bookshop," could be read as "a genuine period bookshop," or "a bookshop *comme il faut*, fallen out of fashion."

As José Pinho, the Alma Mater of Lisbon's Ler Devagar, told me, a bookshop can regenerate the social and economic fabric of an area, because it is the present pure and simple, and a speedy engine of change. That is why we should not be surprised if many bookshops are part of greater social projects for change. I think of those in Latin American cities inspired by Eloísa Cartonera's original shop in Argentina, with its books bound by the unemployed workers who collect paper and cardboard from the streets. I think of La Jícara, a restaurant serving the tastiest of local food wedged between a double bookshop, for both children and adults, which only sells books from independent presses in Oaxaca, Mexico. I think of Housing Works Bookstore Café, which is run exclusively by volunteers and gives all its profits from the sale of books, the renting of space and the cafeteria to help those most in need in New York. They are bookshops that hold out a hand to create human chains. There could be no better metaphor for the book tradition, because we read as much with our hands as with our eyes. I have often heard the same story on my travels. When the time came to change premises, it was the customers, now friends, who offered to help with the move. That human chain uniting the old premises of Auzolán in Pamplona with the new. Or RiverRun's in Portsmouth. Or Robinson Crusoe's in Istanbul. Or Nollegiu's in the Poblenou district of Barcelona.

From at least the time of ancient Rome, bookshops have been spaces for establishing contact, in which textuality becomes more physical than in the lecture theatre or library, because they are so dynamic. In bookshops it is the readers who are most on the move, who bring the copies on display to the counter, interact with the booksellers, take out coins, notes or credit cards and exchange them for books, and who, as they move around, observe what other people are looking for and buying. Books, booksellers and bookshops stay relatively still in comparison with customers who are constantly coming and going, and whose role inside is precisely to circulate. They are travellers in a miniature city whose aim is to provoke the letters—still inside the book—into motion as long as the reading (and its recollection) lasts, because, as Mallarmé wrote: "The book, which is a total expansion of the letter, must derive its mobility from the letter." Nonetheless, the bookshop itself, with or without buyers or browsers, has its own cardiac rhythms. Not only the rhythms involved in unwrapping, arranging, returning and re-ordering. Not only those involved in changes of staff. Bookshelves also enjoy a relationship of conflict with the premises that lodge and partially define them, but do not constitute them. And with their own names, which often alter with successive owners. Inside and out, bookshops are portable and changeable. The reason the Guinness World Record for the Oldest Bookshop in the World is held by the Livraria Bertrand is

because it is the only one that can demonstrate its longevity. Bookshops usually change names when they change hands. At the very least. The oldest in Italy is a case in point: the Libreria Bozzi was founded in 1810 and is still open on a down-at-heel corner in Genoa, but its first owner, a survivor of the French Revolution, was Antonio Beuf; it was not purchased until 1927, by Alberto Colombo, father of the first wife of Mario Bozzi, who gave the establishment, now managed by Tonino Bozzi, its current name. Another good example is the Lello bookshop in Oporto. The establishment was opened under the name Livraria Internacional de Ernesto Chardron, on rua Dos Clérigos; in 1881, José Pinto set it up on rua do Almada; thirteen years later it was sold on by Mathieux Lugan to José Lello and his brother António, who renamed it Sociedade José Pinto Sousa Lello &

Irmão. And if those were not changes enough, after the building of the present edifice—a neo-Gothic and art-deco hybrid—in 1906, the bookshop was given its definitive name in 1919: Livraria Lello & Irmão. An article by Enrique Vila-Matas still hangs in a corner of the shop, where he describes it as the most beautiful bookshop in the world. The card I retain from my 2002 visit is made from elegant, slightly crinkly paper, with the logo and address printed in purple ink. Under the logo it says: "Livraria Lello." "Prólogo Livreiros, S.A." is the name of the company that runs it.

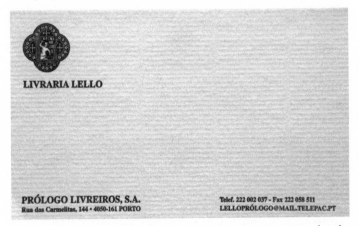

LIVRARIA LELLO

PRÓLOGO LIVREIROS, S.A.
Rua das Carmelitas, 144 • 4050-161 PORTO

Telef. 222 002 037 - Fax 222 058 511
LELLOPRÓLOGO@MAIL.TELEPAC.PT

Another internationally renowned, contemporary bookshop, the Luxemburg, in Turin, tells a similar story: although it was founded in 1872—if we accept that changes of owner, premises and even name do not destroy a bookshop's identity—like Ávila, it had a different name for most of its existence. Owned by Francesco Casanova, an important Piedmontese publisher, the Libreria Casanova was a pre-eminent cultural centre in the final decades of the nineteenth century and the first of the twentieth. The Neapolitan chronicler Matilde Serao, the decadent Antonio Fogarrazo and the creator of *verismo*, Giovanni Verga, were some of its habitués.

Casanova forged a close friendship with Edmondo De Amicis, whose *Gli azzurri e i rossi* he published in 1897. If the premises succeeded in catching the spirit of the times under his management, when the project was taken over in 1963 by activist and writer Angelo Pezzana, who renamed the bookshop Hellas, the new owner also knew how to connect with the times. Given that he was the founder of Fuori!, Italy's first gay rights group, it is hardly surprising that on February 12, 1972 the bookshop launched the countercultural, psychedelic magazine *Tampax* that later engendered another, *Zombie International*. Together with Fernanda Pivano, the great promoter of American literature in Italy, Allen Ginsberg visited the bookshop five years earlier and gave a reading in the basement. When Ginsberg returned to Turin in 1992, he read a continuation of "Hum Bom!," the poem he had begun in 1971, with Bush and Saddam as characters (I'm listening to it on YouTube as I write: an echo of the beat that bookshop had in the 1970s). Pezzana changed the name of the shop again in 1975: Luxemburg Libreria Internazionale. It continued its political and cultural activity: it was behind the inception of the International Gay Association, the Italy-Israel Foundation and the creation of the Turin Book Fair. Under wooden stairs at the back of the first floor, the bookseller's small office is decorated with Italian and Israeli flags, and the Jewish section is almost as well stocked as the international magazines section in the entrance or books in other European languages on the floor above. There is a black-and-white photo of the beat poet and a yellowing press cutting testifying to his visit. One glass cabinet displays invoices and orders made by Francesco Casanova. Pezzana himself, his spectacles teetering on the last millimetre of his nose, takes my money for a copy of Alessandro Baricco's latest novel, which I am buying as a present for Marilena. Access to the basement is shut off.

A Bertrand Livreiros catalogue has been preserved from 1775, the year of the Lisbon earthquake. In it, the French brothers list almost two thousand titles, a third of which are history books, a third sciences and art, and a third law, theology and literature. The majority are written in French and were published in Paris. Many Italian and French booksellers in the Portuguese capital resumed their activities a few months after the earthquake, and although we do not have Bertrand Livreiros catalogues from those years, order forms for titles sent to the Holy Office and to the censorship body that took over its role do exist. They bought the bookshop's definitive premises, in what was then called rua das Portas de Santa Catarina, in one of the public auctions of land devastated by the 1775 earthquake. It remained a family firm until 1876, the year it was sold on by the last direct descendant, João Augusto Bertrand Martin, to the firm Carvalho & Cia. It has since become one of the many brand-name commercial enterprises to incorporate the date 1732 into the initial B, so nobody could question its antiquity.

On the card Pezzana gave me before we said goodbye, it says: "*Fondata nel 1872.*"

IV
Shakespeare and Companies

I shall begin this chapter with a quotation from *L'Histoire par le théâtre* (1865) by Théodore Muret, recorded by Walter Benjamin in his unfinished *The Arcades Project*:

> There were, first, a great many milliners, who worked on large stools facing outwards, without even a window to separate them; and their spirited expressions were, for many strollers, no small part of the place's attractions. And then the Galeries de Bois were the centre of the new book trade.

The association between weaving and writing, between thread and text, between seamstress and artist, is a constant in the history of literature and art. The attraction to artisans and their bodies is related, in Muret's lines, to cultural consumption. He emphasizes the absence of glass, in an era when all bookshops begin to have windows, a transparent display of the merchandise they share with toy or clothes shops. When Zweig describes the return of Jakob Mendel to Vienna, after being interned for two years in a concentration camp, he refers to "window displays of books" in the city, because that is how the inner experience of bookshops is projected outwards, and with it, the exuberance of urban cultural life. The following jotting by Benjamin surely derives from the association of ideas:

> Julius Rodenberg on the small reading room in the Passage de l'Opéra: What a cheerful air this small, half-darkened room has

in my memory, with its high bookshelves, green tables, its red-haired *garçon* (a great lover of books, who was always reading books rather than taking them to others), its German newspapers which every morning gladdened the heart of the German abroad (with the exception of the *Kölnische Zeitung* which on average made an appearance only once every ten days). And when there is any news in Paris, it is here that one can receive it.

Salons, reading rooms, athenaeums, cafés or bookshops act as second homes and political spaces for the exchange of information, as one can see in *The Traveller of the Century* by Andrés Neuman, who also described bookshops as momentary homes. Local and foreign presses enter a dialogue in the *extraterritorial* brains of travellers and exiles, who move from one European capital to another as the Grand Tour dies out. Europe becomes a great space where books flow thanks to their industrial production, which is accompanied by proliferating bookshop chains, the promotion of serial fiction as the main form of commercial novel, an exponential increase in literacy and the transformation of the Continent into a vast tangle of railway tracks. In parallel, the institutions that look after the production and sale of books become stronger. In Germany, for example—as Svend Dahl reminds us—the Association of Booksellers was created in 1825, and twenty-three years later it succeeded in getting censorship abolished and, in 1870, in establishing a norm for the whole country that meant that an author's copyright remained in place for thirty years after his death. By then a system of commission and intermediary wholesalers was in place. Like other consumer goods, books are also subject to the rules of labour legislation, competition, publicity or scandal-mongering.

It was no coincidence that the two major literary scandals of the nineteenth century took place simultaneously in Paris (with apologies to Oscar Wilde, who died, poverty-stricken, in the

French capital). The 1857 prosecutions for offences against public morality and propriety brought against Charles Baudelaire, for his masterpiece, *Les Fleurs du mal*, and against Gustave Flaubert, for his work of genius, *Madame Bovary*, constitute a perfect nexus of controversy with which to illustrate the changes that were taking place in the book industry and the history of literature. Possible answers to questions like: to what extent is a writer responsible for what he writes? And what if it is fiction? Is censorship legitimate in a democratic society? To what degree can a book really influence individuals? What is a publisher's legal relationship with his books? And the printer's, distributor's and bookseller's? These are questions with distinguished precedents. After being denounced by his parish priest, Diderot was prosecuted in 1749 for his *Letter on the Blind* and imprisoned in the fortress of Vincennes until associated booksellers managed to get him released, arguing that if the *Encyclopédie* project continued in abeyance, the nation's industry would be the main victim of the damage. When *El origen del narrador*, which brought together the proceedings of both trials, was published, Daniel Link astutely reinterpreted the volume's title: "Above all it concerns the (modern) notion of author: his simultaneous appearance and disappearance from the scene (of the crime) and the way in which (penal and ethical) responsibility allows specific statements to be related to specific proper names." Baudelaire lost his case (a fine and the suppression of six poems); Flaubert won his. The proceedings reveal that the main protagonist of both trials was the prosecutor Ernest Pinard. Strangely enough, it was in the case that he lost that he showed himself to be an excellent literary critic. We owe to him the interpretation of the novel that still remains in vogue today. Every reader is a critic, but only those who make their opinions about their reading in some way public become literary critics. Pinard was one of the latter, and rightly so, as can be seen from the proceedings.

The poet spent his whole life wanting to write "a history of *Les Fleurs du mal,*" in order to demonstrate that his book was "deeply moral" (although it had been found guilty of immorality). What happened to the book *physically*? Poulet-Masset, its publisher, continued to sell the unexpurgated edition at double the price, and even sold mutilated copies with pages missing. In 1858 he brought out a second edition, now complete again, which sold out in a few months. Unlike Wilde's, which was a genuine tragedy, the scandals provoked by Flaubert and Baudelaire had no serious repercussions. However, they still frame the reading of both masterpieces in the present century—and of the books that followed.

Because of its social impact, the reading of literature is conditioned by countless critical and micro-critical agents. That a critic could be a prosecutor, and that we can pursue the process through the texts he wrote, is extraordinary, as much as if a bookseller were to reveal himself in a similar way. Nonetheless, the two most important Paris booksellers of the first half of the twentieth century—perhaps in the world and of the century—did publish memoirs that enable us to glimpse how key bookshops functioned critically and related to the culture in general. A parallel reading of *Rue de l'Odéon* by Adrienne Monnier and *Shakespeare and Company* by Sylvia Beach allows us to speak of twin projects. Even, by chance, in their initial financing, because Monnier was able to open La Maison des Amis des Livres in 1915 thanks to compensation her father received (as a result of a railway accident) and Beach's mother lent her all her savings so she could invest them in a business that opened its doors at a nearby address in 1919 and moved to l'Odéon two years later. The most important aspect of the trade for both women was the opportunity it afforded to mix with writers who were their customers and who also became their friends. Most of their respective books are dedicated to celebrating their distinguished visitors: Walter Benjamin, André Breton, Paul Valéry, Jules Romain or Léon-Paul Fargue, amongst others, in the case of La Maison des Amis des Livres; Ernest Hemingway, F. Scott Fitzgerald, Jean Prévost, André Gide, James Joyce or Valéry Larbaud, in the case of Shakespeare and Company. That is, if such a division is even possible, because visiting the rue de l'Odéon meant paying a visit to both bookshops and the clientele and friendships of the two booksellers meshed as much in their cultural activities as in their personal lives. While Monnier maintains a degree of balance and devotes similar amounts of space to all her favourite authors, Beach comes down overwhelmingly on the side of Joyce, whom she considered to be "the greatest writer of our era" even before she met him. The entire Joyce family linked up with Shakespeare and Company from the start: youngsters Giorgio and Lucia carried boxes when the bookshop moved from its original premises on rue Dupuytren to its definitive base on l'Odéon, which acted as post office and bank for the whole family, and later Lucia was the lover of Samuel Beckett, her father's assistant, and of Myrsine Moschos, who helped Beach in her bookshop. The story behind the publication of *Ulysses* is the central thread of her book and its author's personality permeates the text, for good or for evil, like a cloud of black-and-white butterflies. I do not think it is pure chance that that book and author are key: *literary bookshops* shape their discourse by creating a sophisticated taste that prefers *difficulty*. As Pierre Bourdieu says in *Distinction: A Social Critique of the Judgement of Taste*: "The whole language of aesthetics is contained in a fundamental refusal of the *facile*, in all the meanings which bourgeois ethics and aesthetics give to the word."

Monnier talks of "the beautiful visits: by authors and well-read fans," Beach of the "pilgrims" who come from the United States, attracted by the aura given the city by the presence of Picasso, Pound or Stravinsky. In fact, she becomes a genuine "tourist guide" when visitors like Sherwood Anderson—one of many—ask her to take them to the residence of Gertrude Stein, and she documents such activity in her pilgrims' sanctuary thanks to the collaborations of Man Ray, whose photographs festoon the establishment. Both places were also lending libraries (in *A Moveable Feast*, Hemingway comments that there was no money to buy books in those days). And Shakespeare and Company also had a guest bedroom. So they acted as art gallery, library and hotel. And embassy: Beach boasts about buying the biggest United States flag in the whole of Paris. And cultural centre: readings and lectures were given periodically in both, and La Maison was home to the first public performance of "Socrate" by Erik Satie in 1919, as well as the first reading of *Ulysses* two years later. Music and literature that were *difficult* and *distinguished*.

Beach decided to keep the bookshop open during the Occupation, but her nationality and Jewish friendships came to the attention of the Nazis. One day, in 1941, "a high-ranking German officer" turned up and in "perfect English" told her that he wanted to buy the copy of *Finnegans Wake* that was in the window. She refused to hand it over. He came back a fortnight later and threatened her. The intellectual decided to close her business and store all its material in a flat in the same building, above which she herself was living. She spent six months in an internment camp. She remained in hiding on her return to Paris: "I visited the rue de l'Odéon daily, secretly, and heard the latest news of Adrienne's bookshop, saw the latest volume of the clandestine Éditions de Minuit." Hemingway was the soldier in the Allied Armies who, in 1944, liberated the street with the mythical bookshops (and then he went to the bar at the Ritz to "liberate" that as well). La Maison continued to be open until 1951, four years before the death of Monnier, who committed suicide after eight months of hearing noises inside her head.

During those decades, Léon-Paul Fargue was the bridge between that Anglo-Saxon-French Paris and Latin American

Paris. Alejo Carpentier describes him as astonishingly erudite and a brilliant poet, always dressed in navy blue, the ultimate wanderer in the night addicted to the metropolis and averse to travelling. Despite his random urban itineraries and lack of punctuality, he was apparently faithful to the Lipp beer house, the Café de Flore—where he would meet up with Picasso—the rue de l'Odéon and Elvira de Alvear's house, where he hobnobbed with Arturo Uslar Pietri and Miguel Ángel Asturias. Another fetish poet and bridge between the two shores was Paul Valéry, whom Victoria Ocampo met in 1928, a crucial visit since she was in the process of preparing the great project of her life, the magazine *Sur*, the first issue of which would be published three years later. Over several months she got to know philosophers, writers and plastic artists. She visited the Russian Lev Shestov in the company of José Ortega y Gasset. She didn't survive her encounter with Pierre Drieu La Rochelle unscathed: they escaped to London embroiled in an adulterous passion. After meeting Monnier and Beach, who introduced her to the work of Virginia Woolf, Ocampo crossed the Channel again in order to meet her in 1934 and returned once more in 1939, accompanied by Gisèle Freund, who took photographs of Woolf that were to become more famous than the ones Man Ray took of Ocampo. The bookselling couple also introduced her to Valéry Larbaud. And Monnier drank tea more than once in the house that Alfonso Reyes and his wife rented in Paris during the previous decade. Nevertheless, to judge by their articles, letters and books, none of these Latin American names resonated in the memories of the Parisian pair of booksellers.

Without a doubt both were radically committed to the literature of their time: the owner of Shakespeare and Company risked her economic well-being to publish the masterpiece of one man and the owner of La Maison des Amis des Livres risked hers to publish her own literary magazine, *Le Navire d'Argent*.

However, Monnier had a more visible profile as a critic than Beach and a greater desire to intervene in contemporary debates. Included in her book is a close reading of the poetry of Pierre Reverdy. Beach recounts an after-dinner conversation with Joyce and Jules Benda where Monnier argued about the best contemporary French writers. In terms of the avant-garde, she states: "We were all very conscious that we were heading towards a renaissance." And on the function of a bookshop with regard to the literary present she says:

> It is truly indispensable that a house devoted to books be founded and directed conscientiously by someone who unites with an erudition that is as vast as possible a love for the spirit of what is new, and who, without falling into the wrongheadedness of any kind of snobbery, is ready to assist the new truths and forms.

To keep both majority and minority happy it is necessary to perform genuine feats of organization and, above all, to keep the space to a minimum. La Maison was a small bookshop and, consequently, it is hardly surprising that its offerings were limited. Many of the writers who paid a visit looked to see whether their books were displayed, or gave them to the library, so it is understandable that the circle of friends and supporters influenced what was for sale, especially if the owner was defending them aesthetically in her interventions in the cultural sphere. In this way the bookshop is transformed into an anomalous place, where exceptional works that, according to Mallarmé, didn't find an opening in modern bookshops are both on sale and find subscribers, investors, translators and publishers.

Monnier writes: "And what discoveries are possible in a book shop, through which inevitably pass, amid the innumerable passers-by, the Pleiades, those who among us already

slightly resemble 'great blue persons,' and who, with a smile, justify what we call our greatest expectations." The bookseller, critic and cultural activist includes herself in the elite. Beyond the difficulties they have finding a publisher or even subsisting, they are the best writers of the period. They possess the aura of recognition: they are *recognized* by those who see them in person, because they may not have read them, but they have seen them in photos, as might have happened with the Eiffel Tower. Chateaubriand says in *Memoirs from Beyond the Tomb*:

> I was in a happy mood; my reputation made my life easy: one dreams endlessly in the first intoxication of fame, and one's eyes are first filled with the joys of the light that breaks through; but when this light goes out, it leaves one in darkness; if it persists, the habit of seeing it soon makes one insensitive to it.

The key word is, of course, *reputation*. Another that is equally crucial depends on it, *consecration*. From the birth of modernity, a highly complex literary system has been articulated through sites of consecration: publication by particular houses, praise from specific critics or writers, translation into certain languages, the winning of awards, prizes, important recognition first locally then internationally, knowing the right people and visiting key cafés, salons and bookshops. Paris during the nineteenth century and the first half of the twentieth constituted the world's pre-eminent republic of letters, the centre where a large slice of world literature was legitimized. When Goethe describes a bookshop in his *Italian Journey* he counterposes three cultural influences: the German he carries with him (and the language in which he is writing his book), the English (the much praised English edition of the book he buys) and the Italian (Palladio and the bookshop itself). As Pascale Casanova has reminded us, Goethe spoke in his work about both a *world literature* and a *world market for cultural goods*. He was fully aware that modernity would be based on the transformation of cultural and artistic objects into merchandise that moves in two parallel markets, the symbolic one (the aim of which is prestige and distinction) and the economic one (the goal of which is the earning of profits for work done, that is part craft and part art).

As is the case with most biographies, essays and most cultural critiques, Casanova's *The World Republic of Letters* does not mention the important role of bookshops in an atmosphere of progressively more international literary geopolitics. Shakespeare and Company is referred to once in relation to Joyce, and La Maison des Amis des Livres appears a few pages earlier, in a paragraph on the topic of the writer as a passer-by with no certified fatherland:

This improbable combination of qualities lastingly established Paris, both in France and throughout the world, as the capital of a republic having neither borders nor boundaries, a universal homeland exempt from all profession of patriotism, a kingdom of literature set up in opposition to the ordinary laws of states, a transnational realm whose sole imperatives are those of art and literature: the universal republic of letters. "Here," wrote Henri Michaux with reference to Adrienne Monnier's bookshop, one of the chief places of literary consecration in Paris, "is the homeland of all those free spirits who have not found a homeland." Paris, therefore, became the capital for those who proclaimed themselves to be stateless and above political laws: in a word, artists.

In the titular 1969 article from *Extraterritorial*, George Steiner speaks of post-modern authors like Borges, Beckett or Nabokov, representatives of a "multilingual imagination," of "internalized translation," the inspiration for their remarkable work. Friedrich Nietzsche was impressed by the existence of trilingual bookshops in Turin when he lived there. Further north, in another polyglot frontier city, Trieste, the Librería Antiquaria, in the inter-war period, was the place where the great writers of Trieste, like Umberto Saba, the poet who ran the bookshop, or his friend Italo Svevo, conversed with writers from other countries, like James Joyce. Consequently, changes of abode and language led to a state of artistic extraterritoriality, but as citizens artists continued to be subject to formal laws and as authors to the rules of the game in their respective literary fields. Although writers in Paris could cultivate a fiction of freedom, it was perhaps easier to do that in relation to geopolitics than in relation to the mechanisms of literary consecration. As well as being a bookseller, Monnier was a literary critic: she judged and she reported. Her important role as a consecrator was recognized by her contemporaries: in 1923 she was accused publicly of exerting a powerful influence with

her book recommendations in *Histoire de la littérature française contemporaine* by René Lalou (according to an article in
Les Cahiers Idéalistes she "ignored those books that weren't on
her shelves"). In her defence, the bookseller argued that she
simply stocked titles that were not available in other bookshops, and, by listing them, she shaped a canon.

The Monnier and Beach duo constituted a doubly anti-
institutional site: respectively, opposing the big local legitimizing platforms (daily newspapers, magazines, universities,
government bodies), and as a clandestine cultural consulate,
opposing the big US legitimizing platforms (especially publishers). From Paris they foiled the American censors and made it
possible to publish Joyce's work in New York: an accomplice of
Beach's ferried copies of *Ulysses* from Canada by hiding them in
her trousers. That anti-national oppositional-space emphasis
hardened during the Nazi Occupation, when it became a
bunker for symbolic resistance.

In 1953, Monnier wrote a piece entitled "Memories of
London" in which she recalls her first trip to the English capital
in 1909, when she was seventeen. It is striking that she does not
mention a single bookshop. Perhaps she had not yet found her
vocation, although people tend to reinforce myths about themselves in retrospective accounts. I think there may be a more
straightforward explanation: at the beginning of the last century, it was difficult to find any awareness of belonging to a
tradition. In fact, the strong tradition of conceptually related
independent bookshops (Shakespeare and Companies) was
born in that transition between Library and Bookshop that was
a revelation for Sylvia Beach:

One day at the Bibliothèque Nationale, I noticed that one
of the reviews—Paul Fort's *Vers et prose*, I think it was—
could be purchased at A. Monnier's bookshop, 7 rue de

l'Odéon, Paris VI. I had not heard the name before, nor was the Odéon quarter familiar to me, but suddenly something drew me irresistibly to the spot where such important things in my life were to happen. I crossed the Seine and was soon in the rue de l'Odéon, with its theatre at the end, reminding me somehow of the colonial houses in Princeton. Halfway up the street on the left was a little grey bookshop with "A. Monnier" above the door. I gazed at the exciting books in the window, then, peering into the shop, saw all round the walls shelves containing volumes in the glistening "crystal paper" overcoats that French books wear while waiting, often for a long time, to be taken to the binder's. There were also some interesting portraits of writers here and there. At a table sat a young woman. A. Monnier herself, no doubt. [. . .] "I like America very much," she said. I replied that I liked France very much. And, as our future collaboration proved, we meant it.

The book was published in 1959 and its natural readership was Anglo-Saxon (hence the comparison with Princeton), as she was fully conscious of the fact that her bookshop was an inevitable reference point and the recreation of its origins would be of interest to literary history. Her tale of discovery is a reader's journey and implies that a frontier must be crossed (the Seine) to reach the unknown. Through the shop window (the second frontier), Beach is linked to Goethe's sense of surprise: businesses still existed that didn't bind their bundles of paper, so the reader could choose the binding to match his or her own taste. The desire in the gaze is concerned as much with the (*attractive*) books on display as with the (*interesting*) portraits of writers, which to this day continue to provide the usual bookshop decor. Their alliance would be finally sealed by a statement of tastes that, with the passage of time, was reinterpreted as a declaration of intent. And of love: Monnier and Beach were a couple for almost fifteen years, although their private relationship does not surface in the books they wrote (nor does the fact that they were among the first women booksellers in the world to be completely independent of male power or investment). That alliance was the first stone laid in their myth. Beach knew she was arriving on the scene four years later and positioned herself in a line initiated by La Maison des Amis des Livres. What she could not know when she published her book was that both bookshops were already part of a tradition connecting the lost generation to the beat generation. Moreover, Beach wrote about the former: "I can't think of a generation less deserving of this name."

The second Shakespeare and Company opened its doors at 37 rue de la Bûcherie in 1951, under the name of Le Mistral, and was not renamed after its admired predecessor until 1964, following Sylvia Beach's death. George Whitman was little more than a scruffy Yankee tramp with some army experience when he arrived in Paris. After graduating in 1935 in science and journalism, he spent several years travelling the world, until the

United States' entry into the Second World War led him to a medical clinic in Greenland, north of the Arctic Circle, and later to a military base in Taunton, Massachusetts, where he opened his first, rudimentary bookshop. It was there, after discovering that personnel were needed in France, he volunteered to help in a camp for orphans. But he was attracted to the capital, so moved there and signed up for a course at the Sorbonne. He bought a few English books with the idea of earning a little from a small lending fee, and suddenly saw his rented bedroom invaded by strangers in search of reading matter; he soon made sure that bread and hot soup were available for those who came to his incipient business. That was the communist embryo of his future bookshop.

Because Whitman was always an uncomfortable figure by US standards. He sold banned books in Paris, like *Tropic of Cancer* by Henry Miller, to soldiers from his country. His American dream followed, as Jeremy Mercer notes, the Marxist principle "Give what you can, take what you need"; and he always saw his project as a kind of utopia. From the very first day in Le Mistral he installed a bed, an oven to warm up food, and a lending library for people who could not afford to buy books. The melding of bookshop and hostel continued for decades, Whitman sacrificed his privacy and continually lived with strangers. In sixty years, some hundred thousand people have lodged in Shakespeare and Company in exchange for a few hours' work in the bookshop spent reading and writing, because new and second-hand books live together, and the presence of sofas and armchairs invites you to use the building as if it were a large library. The presiding motto is written above one of the thresholds within the labyrinth: "Be not inhospitable towards strangers lest they be angels in disguise." An amateur poet, Whitman stated on several occasions that his great work was the bookshop: each of its rooms "was like a different chapter of the same novel."

On one of the windows of Shakespeare and Company it says: "City Lights Books." And above the entrance to City Lights in San Francisco, probably hand-painted by Lawrence Ferlinghetti himself on the green background: "Paris. Shakespeare + Co." Twinned with its Parisian prototype, conscious that it walks the same path, after the four years the beat poet spent studying at the Sorbonne, when he befriended Whitman in his rented room full of books and the smell of simmering soup, the mythical bookshop on the West Coast was born a mere two years after he returned in 1953. It immediately became a publishing house, bringing out books by Ferlinghetti and poets like Denise Levertov, Gregory Corso, William Carlos Williams and Allen Ginsberg. The list was not focused on beat poetry, but many of the books emerged from that orbit: from stories by Bukowski to political texts by Noam Chomsky. The publishing house and its publisher entered literary history in autumn 1955, when Ginsberg gave a reading in the city's Six Gallery: Ferlinghetti suggested publishing "Howl." He did so, and it was very soon withdrawn from circulation by the police, who accused an employee of the bookshop and the publisher of fomenting obscenity. The case received lots of media attention and the verdict, found in favour of City Lights, still constitutes a reference point in the legal history of the United States in matters of freedom of expression. "Books not Bombs" greets you in graffiti on paper hanging in the stairwell. The bookshop defines itself on its walls: "A literary meeting place"; "Welcome, have a seat and read."

Public readings and performance have been a constant from the very start in the Parisian and Californian bookshops. In a famous recital at City Lights in 1959, Ginsberg said he had concentrated hard in order to capture a rhythm to write what he was about to declaim and that, from then on, he improvised with the help of something very similar to *divine inspiration*; he also led readings opposite Shakespeare and Company while drunk on red wine. Both shops are committed to agitation and libraries, to hospitality and openness to the new. Both have well-stocked sections of fanzines that continue to be a means of expression for the same counterculture that emerged in the 1950s. Whitman witnessed the events of May 1968 from the balcony of Shakespeare and Company. It is not by chance that both shops have their poetry and reading rooms at the top of their respective buildings, if one bears in mind their vagabond, beatnik, protest character—neo-Romantics, in a word. Constant renewal is guaranteed in the Paris bookshop by dint of the continual flow of young, temporarily bohemian bodies.

As Ken Goffman writes in *Counterculture Through the Ages*, French artistic society at the beginning of the twentieth century linked the search for artistic originality with bohemian life:

> In the first four decades of the twentieth century, this Parisian artistic bohemia really exploded into something that bordered on a mass movement. Literally hundreds of artists,

writers, and world historic characters whose innovative works (and, in some cases, challenging personas) still resonate today, passed through the portals of what literary historian Donald Piece has labelled "The Paris Moment". . . As Dan Franck, author of the historical work Bohemian Paris: Picasso, Modigliani, Matisse and the Birth of Modern Art, wrote, "Paris . . . [had] become the capital of the world. On the pavements, there would no longer be a handful of artists but hundreds, thousands of them. It was an artistic flowering of a richness and quality never to be rivalled.

The saturation of Paris has an end date: 1939. During the Second World War, cultural life in the city was partially frozen, while the territory and intellectual activity of the United States remained intact. Once the 1940s and their political and military myths passed, cracks opened up in the 1950s that allowed an incipient bohemian life in at bebop pace. A first, quantitative broadening-out takes place from the beat to the beatnik world. Ferlinghetti recounts how busloads of beatniks began to draw up outside the doors of City Lights in the 1960s, as part of their pilgrimage in the tracks of Kerouac, Snyder, Burroughs and the rest. But it is the hippy movement that really turns the new version of bohemia into a mass movement, now entirely stripped of the recherché, distinguished impulses of the first dandies. New levels of literacy and sophistication in The West after the Second World War mean not only that a genuinely mass culture emerges, but that several cultural masses can coexist, each with perfectly defined features.

A consensus must be reached and consequently there have to be followers and readers before a literary generation can be canonized. The last two generations in North American literature—the lost and the beat—entered the canon thanks, among many other factors, to the activity of the first Shakespeare and

Company and its interplay with La Maison des Amis des Livres on the rue de l'Odéon; and to City Lights and the other cultural nuclei of the San Francisco Renaissance, the period of cultural splendour the West Coast city enjoyed in the 1950s. It is no coincidence that "renaissance" is a French word.

V
Bookshops Fated to be Political

A poster of Cicciolina, the then porn actress and future Italian politician, with her bright red lips and swooping neckline, and next to it a poster of the neighbouring Baroque district. A healthy supply of new books and magazines from various countries, on stained walls, under useless, fused bulbs. I discovered such contrasts at the start of the century in La Reduta, a bookshop on Bratislava's Palacky Street, close to what was a quiet park, despite the sparks thrown up by passing trams. That feeling of being between two waters, between two historical moments, is shared by those in every place that has been touched by communism. The buyers devoted as much space to Slovakian literature as to Czech, though new books in Slovakian were thicker, as if proudly underlining the state of play within the context of an extremely slow transition.

The whole of Berlin communicates a similar sense of waters parting. Walking from Alexanderplatz down that wide boulevard, built according to a socialist aesthetic, that was once called Stalin Avenue, and later Karl-Marx-Allee, so broad a whole army could parade down it flanked by several tanks, you are surprised that in that site of megalomania, that perfect scenario for political intimidation, so much emphasis is given to culture. First, you find the Teacher's House's huge mural, with its colourful, didactic exalting of the world of work. A little further on, on the left, you see the façade of Kino International that, from 1963, was the venue for the premieres of the DEFA (Deutsche Film A.G.). After that come Café Moskau, Bar

Babette, the D.S.A. Bar, before you reach the Karl Marx Buchhandlung, the old communist bookshop that, since its closure in 2008, has housed a film production company, and on its left the old Rose-Theater. Two years before it shut, the bookshop acted as the set for the end of *The Lives of Others*, a film essentially about reading.

Stasi captain Gerd Wiesler, who signs off his reports as "H.G.W. X.X./7," spends his whole time reading (listening in on) the daily lives of writer Georg Dreyman and his partner, actor Christa-Maria Sieland. At a key moment in the action, the spy removes a book by Bertolt Brecht from Dreyman's library, a narrow strait through which he timidly verges on the dissident. If in this way the book becomes a symbol of dissenting reading, a typewriter smuggled over from the West—since the secret services controlled every typewriter in the German Democratic Republic—stands out as the symbol of oppositional writing. Dreyman, a supporter of the regime, now disillusioned by the persecution of his friends and the infidelity of his girlfriend (who decides to sleep with a military man in order to avoid being ostracised), types out an article on the extremely high level of suicide that the government is keeping quiet to be published in *Der Spiegel*. Wiesler has begun to feel favourably towards the couple and protects them by writing reports that omit any mention of the suspicious activities being carried out in their house. Thanks to him, the typewriter is not found during a raid and Dreyman is spared the consequences of his treachery, although Christa-Maria accidentally dies during an inspection. As his superior intuits—rightly but without proof—that the spy has switched sides, he reduces him in rank to a purely reading role in the postal service: opening suspects' letters, reading the private correspondence of individuals who might be sending information to the enemy or conspiring to overthrow the regime. After the fall of the Wall, the writer gets access to the Stasi archives where he discovers both the informer and his role in events he had not been able to understand previously. He seeks him out, discovering he is now a postman. He goes from one house to the next delivering envelopes sealed in accordance with the right to privacy. He decides not to say anything. Two years later, Wiesler walks past the Karl Marx Buchhandlug and stops when he recognizes Georg Dreyman on a poster advertising the publication of his latest book. He goes in. The book is dedicated to "H.G.W. X.X./7." "Is it a present?" asks the cashier. "No, it's for me," he replies. The film ends with that response, in this bookshop that is now a large office though I recognize its shelves both from the film and my visits in 2005. I take a photograph of the mural of Karl Marx with his purplish, bearded face, tucked away at one end of the premises. Those traces.

In his novel *Europe Central*, William T. Vollmann enters the mind of one of the spies who acted as perpetual readers of the lives of human beings who, in their eyes, were genuinely literary characters. A critical, censorious mind. His responsibilities include control of Akhmatova's movements. Using a metaphor that was turned into reality by the Stalinist apparatus, he writes: "From my point of view, the correct thing to do would have been to erase her from the picture and then blame the Fascists." Alluding to the sending of subversive material that is much more important than the article written by Dreyman in the film, the spy declares, "For instance, had he been left to me, Solzhenitsyn never could have smuggled his poisonous *Gulag Archipelago* to the other side." Vollmann describes the frantic activity of the bookstalls on the Nevsky Prospect, the cultural artery of St Petersburg, in whose Sytin bookshop Lenin bought his books. Together with bookseller Alexandra Komikova, who sent books ordered by revolutionary militants confined to Siberia, he created the Marxist newspaper that the cause needed in order to spread the word. For *The Development of Capitalism in Russia*, Lenin secured a contract for 2,400 copies, and with the accompanying advance he was able to buy the books he needed for his research in Komikova's shop.

With an honesty rare in literary endeavours, Vollmann recognizes *A Tomb for Boris Davidovitch*, by Danilo Kiš, as the model for his work, where political conflict is taken to extremes in dictatorships of the proletariat, being social constructions based on the existence of legions of ordinary readers and on eminently textual negotiations. Banned books, censorship, translations that are authorized or rejected, accusations, confessions, forms, reports: writing. Based on suspicion, born from horror: writing. In the final struggle between Novski the prisoner and Fedukin his torturer, who tries to drag a full confession from Novski's innards, Kiš captures the essence of the relationship between intellectuals and repressors repeated, like racist jokes, in every community suffering from systematic suspicion. As in *The Encyclopedia of the Dead*, the Serbian writer takes Borges as his starting point, but in this instance he does so in order to politicise him, enriching his legacy with a commitment that is absent from the original:

> Novski lengthened the hearing in an attempt to introduce into the document of his confession, the only record that would remain after his death, a few clarifications that might soften his definitive fall from grace and, at the same time, provide a lead for any future researcher, through cleverly fashioned contradictions and exaggerations, to the fact that the entire fabrication of that confession was based on a lie, wormed out of him, of course, by torture. Consequently, he struggled tenaciously for every word, for every formulation. [. . .] In the last instance, I think that both acted for motives that went beyond any narrow, selfish end: Novski struggled in his death, in his fall, to hold on to his dignity, as any revolutionary would; Fedukin was attempting, within his investigation of the fiction and conjectures, to hold on to what was strictly consistent with revolutionary justice and with those who created that, for it was better to sacrifice one man's or a tiny organism's truth, than to put into question on their behalf principles and interests that were much more sublime.

If the Karl Marx was the most emblematic bookshop in East Berlin, Autorenbuchhandlung was and still is the most influential in West Berlin. Charlottenburg was the centre of the federal half in the divided city and the shop is a few steps away from Savignyplatz, close to the street where Walter Benjamin wrote *One Way Street*, the urban manual that—like Italo Calvino's *Invisible Cities*—helps one finds one's bearings in any metropolitan psycho-geography in the world. The bookshop was inaugurated in 1976 by Günter Grass. As if to signal that its mission was not entirely solemn, Ginsberg turned up a few weeks later—and yet again in this bookshoppy book—to re-inaugurate it with a poetic performance. Until the fall of the Wall, it was a focus for debates about communism and democracy, repression and freedom, with invited speakers like Susan Sontag or Jorge Semprún. In the 1990s it concentrated on cultural reunification, paying great attention to and championing literature from East Germany. As its name suggests, its main distinguishing feature was that it was set up by a group of writers, who took it upon themselves to disseminate the German literature they were producing and reading. The bookshop physically resembles Laie in Barcelona, Eterna Cadencia in Buenos Aires and Robinson Crusoe 389 in Istanbul: sober, elegant and classical. It is fitting that the protagonist of Cees Nooteboom's *All Souls' Day*, a novel with obvious European ambitions, buys his books there.

The axis articulating *Europe Central* is Germany and Russia. In Nooteboom's novel we read:

> It was as if those two countries professed mutual nostalgia for each other that could barely be understood by an Atlantic Netherlander, as if that boundless plain that seemed to begin in Berlin exercised a mysterious power of attraction, from which sooner or later something must again emerge, some-

thing that cannot be understood at this point in time but that, despite all appearances to the contrary, would give another twist to European history, as if that huge landmass could thus turn, twist and drop its western edge like a sheet.

Like atomic bombs with lethally similar content, the regimes of Hitler and Stalin exploded simultaneously in two geographical areas condemned to dialogue after Karl Marx, the Prussian Jew, developed his political ideas. When he was in a seminary, the young Stalin sought freedom to read in the bookshop of Zakharyh Chichinadze, afraid that the books he borrowed from the public library might be checked and lead to reprisals. At the time, imperial censors ruled St Petersburg with an iron hand and encouraged the production of *lubki*, the Russian equivalent of chapbooks or pamphlets that exalted the figure of the Tsar, retold great battles or reproduced popular stories (much to the indignation of pre-revolutionary intellectuals who accused them of being reactionary, anti-Semitic and pro-Orthodox) in Moscow—concentrated in Nikolskaya Street and its vicinity. After the 1917 Revolution they were airbrushed from history. The Great Encounter took place in Chichinadze's bookshop, Stalin having access to Marx's texts. In retrospect, the mythomaniac transformed the experience into an adventure. By his account, he and his companions surreptitiously entered Chichinadze's premises and, hard up, took turns copying the banned books, as described by Robert Service in his biography of the genocidal Soviet leader:

Chichinadze was on the side of those who opposed the Russian establishment in Tbilisi. When the seminarians visited his premises, he surely greeted them warmly; and if copying took place, it must have been with his express or implicit permission. The spread of ideas was more important to the metropolitan elite than mere profit. It was a battle the liberals could scarcely help winning. Chichinadze's was a treasure house for the sort of books the youngsters wanted. Josef Dzhughashvili was fond of Victor Hugo's *Ninety-Three*. He was punished for smuggling it into the seminary; and when in 1896 an inspection turned up a copy of Hugo's *Toilers of the Sea*, Rector Gemorgen meted out "a lengthy stay" in the solitary cell. According to his friend Iremashvili, the group also got hold of texts by Marx, Darwin, Plekhanov and Lenin. Stalin recounted this in 1938, claiming that each member paid five kopecks to borrow the first volume of Marx's *Capital* for a fortnight.

When he won power, Stalin developed a convoluted system of controlling texts, thanks partly to these personal experiences, which had made him realize that all censorship has its weak points. Books have always been key elements in maintaining control of power and governments have developed mechanisms for censoring books, just as they have built castles, fortresses and bunkers that—inevitably—are in the end seized or destroyed, as if unaware of the comment by Tacitus: "On the contrary, the standing of persecuted talent grows, and neither foreign kings nor any that operated with similar fury managed to produce anything but dishonour for themselves and glory for them." It was of course with the printing press that countries began to experience serious problems when they tried to curb the traffic in banned books. And it was under modern dictatorships that the greatest political credit was gained from the public burning of books, at the same time as huge amounts of the national budget were allocated to subsidize *organs of reading*.

In the first centuries of the modern era, Spain pioneered not only massive systems for spying on and repressing readers (how else to describe the Holy Inquisition?) but also routes for importing slaves, concentration camps, schemes for re-education and strategies for extermination. It is hardly surprising that Franco's great rhetorical model for his state was Imperial Spain, the National-Catholic paraphernalia of the conquest of America. The Málaga bookseller Francisco Puche has written about the symbols that were counterposed to the Francoist ones:

> All booksellers who suffered Francoist censorship, police persecution, and fascist bomb attacks were marked for ever by this period and have always believed that a bookshop is more than just a business. We picked up the torch from the last man executed by the Inquisition, a bookseller from Córdoba who was condemned in the nineteenth century for introducing books banned by the Church. And this period made it quite clear, once again, that that reflex action dictatorships have of burning books is no coincidence but the product of two incompatible realities. And it also clearly demonstrated how important independent bookshops are as instruments of democracy.

However, one cannot consider the problematic relationship between aristocratic, dictatorial and fascist regimes and the free circulation of written culture from a Manichaean stance that completely exonerates parliamentary democracies, although fortunately many of them do not have to resort to physical punishment or the death penalty. The United States is the prime example of how freedom of expression and reading have been perpetually besieged by mechanisms of control and censorship. From the 1873 Comstock Law, which focused on obscene and lascivious books, to the present proscription of books enforced by thousands of bookshops, educational institutions and libraries for political or religious reasons, or the ways in which the

Treasury Department's Office of Foreign Assets Control boycotts the diffusion of works from Cuba and other regions in the world, one can see the history of North American democracy as an endless round of negotiations in the fragile area of intellectual freedom. In our era, when any sensational news story is broken, book-burning inevitably catapults onto front pages. As Henry Jenkins has shown in *Convergence Culture,* the books that caused the most controversy in the first decade of this century were the Harry Potter series. In 2002, the books were at the centre of more than five hundred different court cases throughout the US. In Alamogordo, New Mexico, the Christ Community Church burnt thirty copies together with Disney films and Eminem CDs because, according to Jack Brock, the church's pastor, they were Satanic masterpieces and instruments for self-education in the black arts.

It was at the end of the 1980s, however, when the publication of *The Satanic Verses* by Salman Rushdie not only illustrated for the umpteenth time the United States' problematic relationship with direct or indirect censorship, but also placed a much more crucial issue on the agenda: the geopolitical migration of threats to freedom of expression. If for half a century these had been concentrated mostly in Eastern Europe and Asia, from the 1990s they would shift to the Arab world, the difference being that changes in economic relationships and, above all, the media meant that domestic or national polemics could no longer be hurriedly buried by the powers-that-be. From *The Satanic Verses* onwards, the damnation of which coincided with the fall of the Wall, the violence in Tiananmen Square and the unstoppable expansion of the Internet, whenever freedom of expression and reading were under attack, the consequences would automatically be global.

In his memoir *Joseph Anton*, Salman Rushdie recounts the details of his case. Publication initially followed its usual course

in the West: he went on the obligatory promotional tours and was Booker Prize shortlisted, while in India the circulation of *The Satanic Verses* slowly ground to a halt after it was spotlighted in *India Today* ("It will necessarily unleash an avalanche of protest") and the decision by two Muslim Members of Parliament to attack the book on personal grounds (without having read it). All that led to a decision to ban the book. As so often happened in the United States, in India this decision fell to the Treasury Department, guided by the Customs Law. Rushdie wrote in protest to Rajiv Gandhi, the prime minister. In their turn, fanatics responded by sending a death threat to Viking Press, the book's publishing house, and another to a place where the writer was scheduled to do a reading. Then the novel was banned in South Africa. An anonymous message was sent to Rushdie's house in London. Then Saudi Arabia and many other Arab countries banned the novel. And the telephone threats started. Copies of *The Satanic Verses* were publicly burnt in Bradford and the next day "WHSmith, the main British chain of bookshops, withdrew the book from its shelves in its four hundred and thirty shops," while in an official press release, they asked not to be considered "censors." The novel won the Whitbread Prize. A mob attacked the US Centre for Information in Islamabad and five people died as a result of the shooting while the crowd shouted, "Rushdie, you're a dead man!" Then there was Ayatollah Khomeini and his fatwa and two body-guards night and day and a farm in a remote corner of Wales and the threat to boycott all Penguin books on sale in the entire Muslim world and getting to number one in the *New York Times'* bestseller list and a lot of bomb threats and a real bomb that exploded in Cody's bookshop in Berkeley that destroyed shelves, the remains of which are preserved as a reminder of barbarism, and multiple death threats to publishers and foreign translators and the Archbishop of Canterbury, and the Pope's solidarity with the injured feelings of the Muslim people and the Declaration of Support by Writers Throughout the World, then Iran broke off diplomatic relations with Britain and many institutions refused to allow events in support of the writer for security reasons and conflicts escalated ("Those small battles between lovers of books seemed like tragedies in an era when the very freedom of literature was being attacked so violently") and the periodic house moves and a false name ("Joseph Anton") and fire bombs in Collet's and Dillon's bookshops in London and in Abbey's in Australia and in four branches of the Penguin chain and the International Rushdie Defence Committee and daily life conditioned, shot through, shaken by the constant shockwaves from security measures and the first anniversary of the book burning in Bradford and the ratification of the fatwa and the murder of the Japanese translator Hitoshi Igarashi and the ratification of the fatwa and the stabbing of the Italian translator Ettore Capriolo and the ratification of the fatwa and the attempted assassination of the Norwegian publisher William Nygaard and the ratification of the fatwa and the death of thirty-seven people in another protest and eleven years in hiding, eleven unable to stroll along the streets, have dinner quietly with friends in a restaurant, or check that his books were properly displayed in a bookshop. And that his books, on the shelves of a bookshop, should blamelessly lead to so many corpses. So very many.

At the core of Rushdie's description lies an awareness that his book belongs to a tradtion of persecuted literature:

> When friends asked what they could do to help, he often pleaded, "Defend the text." The attack was very specific, yet the defence was often a general one, resting on the mighty principle of freedom of speech. He hoped for, he often felt he needed, a more particular defence. Like the quality defence made in the case of other assaulted books, *Lady Chatterley's Lover*, *Ulysses*, *Lolita*; because this was a violent assault not on the novel in general or free speech per se, but on a particular accumulation of words [. . .] and on the intentions and integrity and ability of the writer who had put those words together.

However, unlike its predecessors, which scandalized a world where news was not spread instantly, *The Satanic Verses* fell victim to a new international context. A context in which the pole of Islamic intransigence sends the other pole into a state of extreme tension, the democracies that, in one way or another, are heirs to the French Revolution. However, if we see the

French Revolution as the first definitive step towards modern democracy, we should remember that alongside the massive number of executions and the sacking of the property of the aristocracy, the people accumulated a huge amount of capital in the form of books that they did not really know how to handle. Alberto Manguel, in *A History of Reading*, reminds us how, at the end of the eighteenth century, when an antique book was much cheaper than a new one, French and German collectors benefited from the revolution, purchasing by weight thousands of bibliographic jewels through French intermediaries. As the literacy level of ordinary people was very low, the books that were not sold or destroyed did not find too many readers in the public libraries where they were sent. Nor did the opening of public galleries lead to immediate cultural consumption: the most important consequences of collective education are always long-term. The redistribution of books would bear fruit several generations later. A large number of Islamic countries are now working to consolidate systems for repressing reading in order to ensure a future without plurality, dissension and irony.

In the history of Foyles, the prestigious London bookshop, we find another triangle, two sides of which are in Germany and Russia, through the same dynamic that has been repeated from time immemorial: wars, revolutions, political changes of a radical nature as moments that encourage huge quantities of books to change sides and owners. When Hitler began massively burning books in the 1930s, the first thing that William Foyle thought to do was to send him a telegram offering him a good price for those tons of inflammable printed material. Shortly before that he had sent his daughter Christina, then in her twenties, to Stalinist Russia in search of bargains. The Russian expedition was a success, but not the German sortie: Hitler continued to burn books and had no intention of selling them. Once war broke out and London fell victim to the Nazi bombing raids, the old books from the cellar were mixed with sand and filled the bags that protected the walls of his shop whilst, apparently, Mr Foyle covered the roof with copies of *Mein Kampf.*

There were certainly copies of *My Struggle*, the English edition published by Hurst & Blackett and translated by Edgar Dugdale, a Zionist activist who did so with a view to denouncing Hitler's plans. Unfortunately, both the English and North American publishers (*My Battle*) yielded to the demands of Eher-Verlag, which compelled them to cut out the many xenophobic and anti-Semitic rants in the original. As Antoine Vitkine explains in his history of the book, as soon as it appeared in Britain in 1934, 18,000 copies were sold, but by that time it had been read by Churchill, Roosevelt, Ben-Gurion and Stalin, who had unexpurgated translations made by their intelligence services. *Mein Kampf* not only turned Adolf Hitler into *the* bestselling author of 1930s Germany, and a millionaire thanks to his royalties, it also made him think of himself as a *writer*, which is how he describes himself in the corresponding section of his income tax returns for 1935. There is no doubt that being the country's political leader helped his sales, though the writing myth (prison) and his Messianic will also helped spread the word at a dizzying rate, as did conveniently placed advertisements in the main newspapers of the day. Instead of a typical bookshop launch, Hitler decided on the Bürgerbräukeller to promote the work of his life:

> It is a clumsy and contrived argument, but he convinces his audience. In order to struggle against the shades of Marx, a Nazi Marx is needed or, in other words, Hitler himself, the author of *Mein Kampf.* By presenting himself as a writer, Hitler changes his image and emerges from the mud where he had operated until then. He is no longer simply a beerhouse braggart, a loudmouth, a failed putschist: now he covers

himself with the prestige that comes with letters and appears as a new theorist. When he leaves the room, Hitler's men hand out promotional leaflets advertising the publication of his book and even specifying the price.

His fame as a book burner eclipsed his fame as a book collector: by the time of his death the exterminator had amassed a library of more than 1,500 volumes. After leaving school, in the shift from adolescence to young man that was accompanied by lung problems, Hitler devoted himself to the life of the artist and intellectual, drawing and reading compulsively. He never gave up on the second activity. August Kubizek, his only friend from the Linz years, relates how he used to go to the Popular Educational Society bookshop on Bismarkstrasse, and several lending libraries. He recalls seeing him surrounded by piles of books, especially the Sagas of German Heroes collection.

Some fifteen years later, on the other side of the world, while Hitler was staging his first mass rally and setting the Nazi propaganda machine in motion, another future perpetrator of genocide, Mao Tse-tung, was opening a book-shop and publishing house in Changsa, which he dubbed the Cultural Society of Books. Business was so brisk he soon had six employees, thanks to whom he could spend his time writing political articles that brought him to the attention of leading Chinese intellectuals. He fell in love and married in the same period. In previous years he had worked as a librarian, assistant to Li Dazhao, one of the first Chinese communists, in whose study group he was introduced to the basic texts of Marxism-Leninism. But it was in 1920, when he became a bookseller, that he began to call himself a communist. Forty-six years later he spurred on the Cultural Revolution, one of whose fronts was the burning of books.

As the world's principal communist regime, China supports state chains that open vast bookshops in the country's main cities, oversee public morality and abundantly stock the Studies of Success sections, in order to foment hard work and the surpassing of individual expectation, the basis of collective endeavours. The Shinhua chain is probably the biggest and owns monsters like the Beijing Book Building, at the junction of two metro lines, with 300,000 volumes spread over five floors. The titles selected by the government coexist on its shelves with bestselling literature, school textbooks and some books in English. However, in the University of Military Science, the School of the Central Party, and the University of National Defence bookshops, official output isn't covered by layers of pretence: they publish works of statistics and forecasts written by officers in the People's Army, doctoral theses and studies that reveal the hard core of communist thought, undisguised by the camouflage of official communiqués intended for the foreign press. Fortunately, beneath its glamorous veneer and distinction as being one of the most beautiful bookshops in the world, the Book Worm bookshop in Beijing has, over the last few years, offered its customers banned or dissident books like those by the artist Ai Weiwei.

The last time I went to Venezuela, a very young soldier smelled the twenty-three books I was carrying in my luggage one by one. I asked him whether drugs now travelled inside literature and he gave me a puzzled look before replying that they mixed them with glue, in the binding. He also sniffed the two volumes from the Biblioteca Ayacucho that I had bought in a Librería del Sur, the bookshop chain run by the Ministry for the Popular Power of Culture of the Bolivarian Government of Venezuela. When he finished his inspection, he grasped my iPad, relaxed his tone of voice and asked me if I had bought it in the United States and how much it had cost. Apart from Maiquetía, they have scrutinized the books in my luggage in two other airports—title by title and running a thumb across the pages: in Tel Aviv and in Havana. Israeli spies are very young and are often doing compulsory military service; while holding one of your books they ask you if you are planning to visit Palestine, or if you have been there and brought something back, and who you know in the country, where you will stay or have stayed, why you have come, and transfer the information to a label they stick in your passport that evaluates the level of danger you represent. Venezuelan soldiers dress exactly like their Cuban counterparts and are equally unsophisticated; they are in fact copies of that original Cuban style.

It was in the communist bookshop on calle Carlos III in Havana that future commander and repressor Fidel Castro bought the two key books of his life: *The Communist Manifesto* and *State and Revolution*. When he was in prison, he devoured all manner of reading matter, from Victor Hugo and Zweig to Marx or Weber, volumes that were presents from people who visited him in prison; he had bought many others in the same bookshop on Carlos III. In *Un seguidor de Montaigne mira La Habana* (*A Follower of Montaigne Looks at Havana*), Antonio José Ponte remembers how you could once buy books in Russian on calle Obispo in the old city:

> I found an old photograph in an encyclopedia from the beginning of the century: a street with shops and awnings on both pavements, it looks like a souk, an Arab market seen from on high. Some time ago I wrote it is rather beachy. It begins with the bookshops and ends by opening out into the square and the port. One of the bookshops then sold books in Russian. Soviet ships passed through the port. Obispo was framed by two notices in Cyrillic: the title of a book and the name of a boat.

But it is in *La fiesta vigilada* (*The Fiesta under Surveillance*) where Ponte traces more precisely the tortured topography of Castro's city, capital of "the theme park of the Cold War." He evokes Comandante Guevara in all his complexity: revolutionary soldier and professional photographer, political leader and writer by vocation, a keen reader. "From his military headquarters in La Cabaña," he tells us in one masterly sentence, "Ernesto Guevara managed a magazine, the camp's military band, the army's film unit and the execution squad." The Revolution provoked, and still provokes, wave after wave of revolutionary tourists. At one point in his book, Ponte recalls the experiences of Jean-Paul Sartre and Susan Sontag, his firmness and her doubts, how Nicolás Guillén's chilling words echo in the wake of their footsteps: "Any investigation is counter-revolutionary." In the last section, the narrator moves to Berlin, where he meets his translator, who has just got hold of the Stasi's file on him: "Thanks to a neighbour who spied on his movements, he was able to revisit a day in his life from thirty years ago." That trip allows Ponte to transform his life as a writer under surveillance in Havana into a universal experience.

It was a long journey that brought Che from Buenos Aires to Cuba. And a reverse journey, from north to south, ended in a laundry in El Señor de Malta public hospital, Valle Grande, his corpse before the lens of Freddy Alborta. I met Freddy by chance in his photography shop in La Paz, shortly before his own death, and he told me the story of that other journey: its result, photographs of the illustrious corpse in a glass cabinet alongside rolls of film and frames, on sale like postcards. In one of the best known, Bolivian military officers pose next to the body, as if it were an improvized anatomy lesson, and one touches the stiff body with his index finger, gesturing towards it, demonstrating that myths are also made of flesh, of matter in a constant process of putrefaction.

Would the books of Ernesto Guevara the writer be on sale in the Librería Universal? I don't think so. In the same year that the revolutionary was appointed President of the National Bank and Minister of the Economy, counter-revolutionary Juan Manuel Salvat abandoned the island via Guantánamo. Five years later he and his wife opened what was to become one of the cultural focuses of the exiled community on 8th Street in the city of Miami, with its literary conversations and editions of books in Spanish. In a report by Maye Primera, prompted by the closure of the Librería Universal on June 20, 2013, Salvat declared that the first generation of exiled Cubans, the one that read most, was dying out and the "new generation, our children, although they feel Cuban, have no experience of Cuba, don't have the accoutrements of nationality, and their first language is English, not Spanish." A law of life.

On May 2, 1911 Pedro Henríquez Ureña wrote a letter from the Cuban capital to Alfonso Reyes in which he said: "But don't think for one minute that there are good second-hand or new bookshops here: the Havana bookshops aren't much better than those in Puebla." It is possible that for a Mexican visitor the city's bookshops did not seem anything special at the beginning of the last century, but calle Obispo—in whose Hotel Ambos Mundos Hemingway liked to stay—and the Plaza de Armas were the heart of the book trade, where the citizens of Havana obtained reading supplies in the decades when they were unable to travel. When I visited the island in the last days of 1999, I only bought books on the stalls in the Plaza de Armas, because the state shops had very few titles on offer, and filled up all those square metres of space with dozens and dozens of copies of the same title. In doorways, garages and entrance halls, second-hand books were on sale: people were offloading family heirlooms for a fistful of dollars. But La Casa de las Américas, once the powerful bastion of Latin American culture, displayed only a few volumes by writers who were officially approved. Jorge Edwards, who at the end of the 1960s was a jury member for its prestigious annual prizes, recounted in *Persona non grata* the brutal turn made by the regime at the beginning of the 1970s. The Chilean writer gives many examples of these changes, unfortunately inscribed in the DNA of the very idea of a communist revolution, and very similar to those related by Kiš and Vollmann in their stories about paranoia in the Soviet orbit, though one is particularly telling. The rector of the University of Havana informs him, "We in Cuba don't need critics, because it is very easy to criticize, you can criticize anything, the difficult thing is to build a country and what that needs is creators, builders of society." So much so that they wonder whether to suppress a magazine whose name suddenly seems highly subversive: *Pensamiento Crítico*. And Raúl Castro conspires to subject theoretical studies of Marxism to army control. I read that book, and also *Before Night Falls* by Reinaldo Arenas, in the first days of the turn of the century, part of an archive of the degeneration that had been gathering pace for over thirty years. As if all the work carried out then—which one can imagine, for example, when reading Cortázar's letters—had been drained away and the shelves of Rayuela, the Casa de las Américas' bookshop, were the end result of that draining away.

I can think of few images that are sadder than an almost empty bookshop or the remains of a bonfire on which books have been burnt. In the sixteenth century, the Sorbonne decreed that five hundred books were heretical. At the end of the eighteenth 7,400 titles were listed in the *Index of Banned Books* and, when the revolutionaries took La Bastille, they found a mountain of books that were about to be burnt. In the 1920s the United States Postal Service burnt copies of *Ulysses*. Until the 1960s it was impossible to publish D. H. Lawrence's *Lady Chatterley's Lover* or Henry Miller's *Tropic of Cancer* in Britain and the United States legally without charges of obscenity. In 1930, the Soviet Union banned private publishing and official censorship lasted until Perestroika. Eugenio Pacelli, the future Pius XII, read *Mein Kampf* in 1934 and persuaded Pius XI that it would be better not to include it in the *Index*, to avoid infuriating the Führer. Books were publicly burnt by the most recent dictatorships in Chile and Argentina. Serbian mortars tried to destroy the National Library of Sarajevo. Periodically puritan, Christian and Muslim demonstrators burn books just as they burn flags. The Nazi government destroyed millions of books by Jewish writers at the same time it was exterminating millions of Jewish human beings, homosexuals, political prisoners, gypsies and sick people; though it preserved a few—the rarest or most beautiful—with the intention of putting them on display in a museum of Judaism that would only open its doors after the Final Solution had reached a definitive conclusion. We have often been reminded of how the Nazis in charge of the death camps were fond of classical music; but people rarely consider, on the other hand, how those who designed the biggest systems of control, repression and execution in the contemporary world, who showed themselves to be the most effective censors of books, were also individuals who studied culture, who were writers, *keen* readers. In a word: lovers of bookshops.

VI
An Oriental Bookshop

Where does the West end and the East begin? There is no answer to this question. Perhaps there was in more distant days: in Flaubert's time, maybe, or much earlier, in Marco Polo's, or much, much earlier, in Alexander the Great's. Nevertheless, Western thought in ancient Greece was created in dialogue with philosophies from the other shores of the Mediterranean, so it was already in itself thought that encompassed an abstraction known as *the Oriental*, even though later rereadings tried to efface that. But this chapter must start somewhere, as previous ones have in Athens or in Bratislava, and so we will begin in Budapest, one of those cities—like Venice, like Palermo, like Smyrna—that seem to be adrift between two different waters that are less in contradiction than in conversation.

It was on a summer's day at the beginning of this century that, on one of my strolls through the city, I finally became infatuated with a peculiar hand-painted wooden box: it did not open and hence seemed completely useless. A green wooden cube with filigree decorations was on display alongside other souvenirs on one of the stalls on the banks of the Danube. It clearly had a lid but no keyhole. The stallholder waited a while until she could see I was desperately turning that hermetic object round in my hands, then she came over and whispered, "It is a magic box." A few movements of her fingers laid bare loose pieces in the wooden structure, parts that slid one way and another to reveal a keyhole and, indeed, the crevice where the key was hidden. The device had entranced me. She realized that immediately. The haggling began.

The dichotomy between fixed prices and haggling could be one of the axes that today polarize East and West. Another could be the material and the oral. These are uncontrollable, slippery points of opposition, but they can help us to decide whether categories like "the Western reader" or "the Oriental bookshop" have any meaning. In Marrakesh's Djemaa el-Fna Square, the library is non-material and inaccessible to those who don't know the local languages: the snake charmers, ointment-sellers and, above all, the storytellers, construct an incomprehensible story out of thin air accompanied by hypnotic gestures, illustrations using human bodies or drawn maps. In *The Voices of Marrakesh*, Canetti links this lack of understanding with a degree of nostalgia for artisanal ways of life that have died out in Europe, ones that give more credence to the oral transmission of knowledge. No doubt wisdom is the greatest value in the oral traditions that flow into that dusty, rather caravanserai-like square that every afternoon turns into a huge, informal, steaming, open-air diner. However, to idealize it is to return to an Orientalist mentality, to the clichés and simplifications in relation to the Arab and Asian worlds that we so-called *Westerners* like to traffic in, just like that image of an Egyptian bookseller I photographed in a small village on the shores of the Red Sea. After all, the Arab and Asian worlds are worlds of calligraphy and books with ancient, powerful texts that are closed to us unless we partially betray them through translation.

Tangier's proximity to Europe meant that it soon began to be Orientalized by European writers and painters, particularly the French. Delacroix was the first to turn the Moroccan city into a huge, abstract landscape. His repertoire of *djellabas* and horses, young boys and carpets, against a simple white architectural backdrop where a glassy sea often appears comprises the clichés that will be repeated time and again in the representation of North Africa. Eighty years later, as part of the same

tradition, Matisse gave geometrical form to the city and its inhabitants: he modernized it. Among the Spanish painters, Mariano Fortuny, Antonio Fuentes and José Hernández gradually added different shades to that pictorial landscape. The latter, a member of the city's Hispanic community, exhibited in the Librairie des Colonnes, perhaps the city's most important cultural centre over the last sixty years. It was also where the writer Ángel Vázquez worked. He won the Planeta Prize in 1962 and fifteen years later published his great novel about the city, *La vida perra de Juanita Narboni*. People tend to remember the roll-call of American and French artists who made the International City one of the key focal points of twentieth-century culture, but nonconformist figures from many parts of the globe hovered around them, like the Spaniards I have mentioned, or Claudio Bravo, the Chilean hyper-realist painter who lived in Tangier from 1972 until his death in 2011, or the

Moroccan artists who participated in the creation of the myth, like Mohamed Hamri the painter, or the writers Mohamed Choukri, Abdeslam Boulaich, Larbi Layachi, Mohammed Mrabet or Ahmed Yacoubi.

The official narrative of what could be called the *Tangier myth* places its beginning in 1947, when Paul Bowles arrived in the city. The following year his wife Jane joined him. Tennessee Williams, Truman Capote, Jean Genet, William Burroughs (and the rest of the beat generation) or Juan Goytisolo would appear later. Beyond numerous parties in private homes and certain cafés that became daily points of encounter, two main meeting places emerged for the motley band of artists and numerous other characters who came and went, tycoons and adventurers, dilettantes and musicians interested in African rhythms, actors like Hungary's Paul Lukas (who appeared with Elvis Presley in *Fun in Acapulco* and in the version of *Lord Jim* directed by Richard Brooks, and who died in Tangier while he was looking for a place to spend the last years of his life), film directors like Bernardo Bertolucci and rock groups like the Rolling Stones. These two meeting places were, on the one hand, Paul Bowles himself, who became a tourist attraction similar to Gertrude Stein or Sylvia Beach in Paris in the inter-war years; on the other, the Librairie des Colonnes, founded in the period when the Bowleses settled in Tangier, and which has survived them both.

The Belgian couple, Robert—architect and archaeologist, friend of Genet, André Gide and Malcolm Forbes—and Yvonne Gerofi, a librarian by training, took the helm of the Librairie des Colonnes from its founding in 1949, with the indispensable collaboration of her sister, Isabelle. It was Gallimard, the owner of the business, who offered them the post. Their marriage was paper-thin. The couple had gotten together for convenience as both were homosexuals and Tangier at that time was the ideal place for the Bowleses' kind of family set-up. While the Gerofi sisters assumed control of the bookshop, becoming genuine celebrities in the cultural sphere, Robert devoted himself to design and architecture. Among other projects, he was responsible for reshaping the Arab palace where Forbes, the publisher and owner of the famous magazine, housed his collection of 100,000 lead toy soldiers. In a Magnum Agency photograph, an old man appears, looking at the camera, holding a white jacket and white hat, as "manager of the Forbes Estate." The writers' correspondence suggests the Gerofis and the Bowleses were in close contact. As far as Paul was concerned, the Gerofis' constant presence went without saying: they were as much a part of the daily scene as the Zoco Chico or the Straits of Gibraltar. On the other hand, when she wasn't being her nurse, Yvonne was a close friend of Jane's, since Jane depended on her during the long periods when she was psychologically unstable. On January 17, 1968 Jane walked into the Librairie des Colonnes not recognizing anyone, her mind a complete blank, and asked to borrow two dirhams; then she took two books, and, despite the entreaties of her servant Aicha, left without paying for them.

Whenever Marguerite Yourcenar passed through Tangier, she would drop in at the bookshop to greet her friend Robert, and whenever an American writer—like Gore Vidal—or a European intellectual—like Paul Morand—or an Arab—like Amin Maalouf—visited the white city, they inevitably ended up between shelves that, over time, came to stock a varied selection of Arab, English and Spanish titles, as well as an unexpected wealth of French books. Not for nothing was it a bastion of anti-Francoist resistance that encouraged publications by and organized meetings of exiles. The most renowned Spanish writer linked to the Librairie des Colonnes is Juan Goytisolo, who began to inhabit Arab culture in the mid-1960s, in that very city. As soon as he arrived, he wrote to Monique Lange, as we read in *In Realms of Strife*: "I feel happy, I walk around for ten hours a day, I'm seeing Haro and his wife, I'm not going to bed with anyone and I look at Spain from afar, full of intellectual excitement." He writes of the inspiration for *Count Julian*: "The idea I'm working on is based on the vision of the Spanish coast from Tangier: I want to start off with this image and write something beautiful that will go beyond anything I have written so far." He takes detailed notes, sketches out ideas and reads Golden Age literature profusely in his rented room. Although he later decides to settle in Marrakesh, Goytisolo will spend most of his summers in Tangier and become a supporter of its most important bookshop. In one of his most recent novels, *A Cock-Eyed Comedy*, where he gives a turn of the screw to the camouflaged homosexual tradition in Spanish literature, the colourful père de Trennes declares:

> Do you know whether Genet still stays at the Minzeh or has he set up in Larache? I've heard great things about an autobiography by one Choukri, translated into English by Paul Bowles. Have you read it? As soon as we arrive I'll track it down in the Librairie des Colonnes. You're a friend of the Gerofi sisters, I suppose? Who doesn't know the Gerofi sisters in Tangier? What! You've never heard of them? *Pas possible!* Doesn't an honorary Tangerine like you go to their bookshop? Allow me to say I don't believe you. They're the engine driving the city's intellectual life!

The case of Eduardo Haro Ibars is less well known, but perhaps more emblematic given his affinity with bisexuality, drugs and the destructive inertia that permeated the intellectual climate in Tangier. The son of exiles, born in Tangier in 1948, he infiltrated the beat circle as an adolescent who accompanied Ginsberg and Corso on their nocturnal wanderings. "I grew up rather a roamer, between Madrid, Paris and Tangier," he wrote; but it was surely the spatial vector of Tangier–Madrid that really marked his life, because he took to the Spanish capital a nonconformist injection of beat which, as a militant homosexual, he used to nourish *la movida*, writing poems and songs and experimenting with all kinds of hallucinogenic drugs. In the spring of 1969, after four months in prison with Leopoldo María Panero, he returned to the family home in Tangier. On another occasion, he boarded a night train that took him to Algeciras, crossed the straits, lodged in the house of Joseph McPhillips—a friend of the Bowleses'— and was helped by the Gerofi ladies, who let him do a few jobs in their bookshop. He defined himself as "homosexual, drug addict, delinquent and poet."

He died of complication from AIDS at the age of forty.

The word is electric star in the fog

Bookshops tend to survive both their owners and the writers who feed their mythology. After the Gerofis, between 1973 and 1998, the business was looked after by Rachel Muyal. As we read in *Mis anos en la librairie des colonnes*, as a Tangerine and neighbour of the bookshop from 1949, she brought to the cosmopolitanism she inherited her added interest in the Moroccan nature of Tangier:

> A person who honoured me with his visits was Si Ahmed Balafrej. He liked to browse through the interior design and architectural magazines. Si Adelkebir el Fassi, a resistance hero, used to accompany him. It was in the course of one of their conversations that Si Ahmed looked me in the eye and said, "Only God knows that I have done everything to ensure Tangier preserves its special status whilst remaining part of the Kingdom of Morocco."

Like other great booksellers who have appeared or will appear in these pages, Muyal lived within a stone's throw of the shop and often organized cocktail parties and fiestas linked to the launch of books or cultural events, and like them she also became a point of reference, an ambassador, a link: on a weekly basis three or four people would ask her to put them in contact with Paul Bowles, who didn't have a telephone; she'd ask for appointments via messengers and he almost always granted them.

Later, Pierre Bergé and Simon-Pierre Hamelin arrived, and with them, the magazine *Nejma*, which has devoted its pages to the memory of the international myth, to that map where so many Moroccan writers found paths to translation and recognition outside Tangier. The Straits of Gibraltar have always been a place of transit between Africa and Europe, so it is only natural that the bookshop should have played a privileged role in the cultural communication between the two shores. Muyal declared in a lecture she gave to the city's Rotary Club:

> I could feel myself in the centre of the city and even in the centre of the world in that mythical place that is the Librairie des Colonnes. That is why, I told myself, it was absolutely necessary to make the institution participate in the cultural movement in Tangier, a city that symbolizes better than any other in the world the meeting of two continents, two seas, two poles: East and West and also three cultures and three religions constituting a single, homogeneous population.

I still have the hand-woven paper card of the Librairie Papeterie run by Mademoiselle El Ghazzali Amal in Marrakesh, on which she has proudly embossed: "Since 1956," and I remember how disappointed I was by the scant number of books on sale and by the fact they were all written in Arabic. The Librairie des Colonnes, on the other hand, can only enthuse a European writer, because it is like any great bookshop, but just happens to be on the shores of Africa and possesses all the necessary local colour. It sells fixed-price books in French, English and Spanish, without the option of haggling that is amusing initially but soon becomes trying and wearisome. The same is the case with the other two Moroccan bookshops I have gotten to know recently: the Ahmed Chatr, also in Marrakesh, and in particular the Carrefour des Livres in Casablanca, with its stridently coloured canvases and large stock of titles in Arabic and French (it has direct links with the Librairie des Colonnes, since it sells the same small white-and-tangerine books from the Khar Bladna house that I have been collecting over the years). You feel at ease. I have rarely had such a feeling of suffocation as in that other Marrakesh bookshop, which was dedicated exclusively to religious books, entirely in Arabic, without a single breathing space. We travel to discover but also to recognize. Only a balance between those two activities can give us the pleasure we are seeking. Bookshops are almost always a sure-fire bet in that respect: their structures are soothing, because they always seem familiar; intuitively we understand the orderliness, the layout, what they have to offer, but we need at least one section where we recognize an alphabet we can read, an area of illustrated books we can leaf through, a scattering of information that, in its precision—or simply by chance—we can decipher.

That was exactly what occurred in the Book Bazaar in Istanbul: among thousands of incomprehensible covers I found a volume about Turkish travellers published in English and illustrated with photographs, *Through the Eyes of Turkish*

Travellers, Seven Seas and Five Continents, by Alpay Kabacali, in an exquisite cased edition published by Toprakbank. As I needed that piece of the puzzle—accounts by Turkish travellers—in my historical travels collection, I was determined to buy it. Right away I was reminded of the seller of magic boxes on that Budapest street stall, where I had gone day after day, keeping firmly to my offer—a third of the asking price—until she yielded on the last day with a feigned smile of resignation. I bought two, to give to my brothers. The very minute she handed them over, wrapped in grey paper, an American tourist, holding an identical box, was asking how much they cost. The lady doubled the initial price. Without objecting, her customer also asked for two and put his hand in his pocket and, highly amused, with a wink that begged me not to say a word, she agreed to a sale six times more expensive than mine. So I asked the Turkish shop assistant who was listening to the radio behind the counter for the price of the cased volume, though it turned out that he was only keeping an eye on the merchandise, because he immediately shouted to a clean-shaven, middle-aged man who looked me in the eye and said it cost forty dollars. I think twenty-five would be a fairer price, I replied. He shrugged and went back by the route he had come.

He had come round one of the corners of the Sahaflar Carsisi that is Turkish for "Book Bazaar." Located in an old courtyard, boxed between the Beyazit Mosque and the Fesciler entrance to the Great Bazaar, close to Istanbul University, it occupies approximately the same number of square metres enjoyed by the Chartoprateia that was Byzantium's market for paper and books. Perhaps because a bust of Ibrahim Müteferrika sat in the centre of the courtyard, accompanied by the titles of the first seventeen books published in Turkish thanks to the printing house he ran from the beginning of the eighteenth century, I thought I might be able to secure the anthology of travel writers employing the same tactic I had used in Budapest. Because Müteferrika hailed from Transylvania and we don't how he came to Constantinople or why he converted to Islam, in my eyes his Turkish journey was linked with my incursions in the Balkans and along the Danube. I soon got into the habit of going there every day and upping my offer by five dollars on each visit.

I also adopted the habit of reading in the afternoons on the terrace of the Café Pierre Loti, with its views over the Marmara, and strolling at nightfall along the Istiklal Caddesi, or Independence Avenue, the other great book centre in the city. Like Buda and Pest, the two banks of Istanbul separated by the Galata Bridge have their own idiosyncratic character which could be summed up by the two different focuses on writing: the Bazaar and the Avenue. Merchants from Venice and Genoa established themselves around the latter; there are beautiful arcades and bookshops with the price of each book printed on a white label stuck on the back cover. I looked in vain for the travellers' anthology in places like

Robinson Crusoe 389, where, conversely, I bought two books by Juan Goytisolo translated into Turkish. The photographs included in the edition of *Ottoman Istanbul* do not include any of the old or modern bookshops, because they have never featured in travel literature or cultural history. I searched for a bibliography on the Armenian genocide and, at the end of the avenue that looks over the Galata Tower, I finally found a bookseller who spoke perfect English—with a London accent—who referred me to the two volumes of *The History of the Ottoman Empire and Modern Turkey* by Stanford J. Shaw and Ezel Kural Shaw. Their index of topics left no room for doubt: "Armenian nationalism, terrorism; Armenian revolt; the Armenian question; war with Turkish nationalists." Equally shocking but less comprehensible is the fact that the historical summary offered by *Lonely Planet: Turkey* also avoids mention of those systematic massacres that took the lives of a million people, the first of the genocides of the twentieth century.

In premises close to the bust of the first Turkish printer—who was Hungarian—I had several conversations with a bookseller who spoke good English and who—as the days went by—became less and less wary. Orhan Pamuk, who had just won the Nobel Prize, was, so he said, a mediocre writer who had made the most of his foreign contacts. And the Armenian genocide was an episode in history that really did not deserve that name, because one should separate out facts from propaganda. I cannot work out if his name was Burak Türkmenoğlu or Rasim Yüksel, because I have kept his card as well as one from the middle-aged, always freshly shaven man, who on the day I took the night bus to Athens sold me the blue-cased volume for forty dollars. However, I do remember very clearly the way his eyes shone like silver paper reflecting flames in the half-dark of his premises.

There is an abundance of denial literature in Turkey, as there is of anti-Semitic material in Egypt or anti-Islamic books in Israel. In the Madbouly bookshop on Talaat Harb Square in Cairo I saw three copies of *The Protocols of the Elders of Zion*, though they also had the complete works of Naguib Mahfouz on display, the only Egyptian writer who emulated Stein or Bowles and transformed himself into a tourist attraction in his lifetime, as a regular customer at the Fishawi or Café of Mirrors. In Sefer Ve Sefel in Jerusalem, which was founded in 1975 with the idea of offering books in English and had to close down during the intifada, or in Tamir Books, on the same Jaffa Road, where they only sell books in Hebrew, and all the different political and historical tendencies, including the indefensible kind, coexist: generalist bookshops tend to be a microcosm of the wider society: radical minorities are represented on shelves that are also in themselves minimal. But I went to fewer book-shops in Jerusalem than in Tel Aviv, a city less obsessed by reli-gions and thus more tolerant, and the bookshop I visited daily during my stay in Cairo was the one in the American University that is apolitical and secular, and deliberately so. There I bought one of the most beautiful books I have ever given as a present: *Contemporary Arabic Calligraphy* by Nihad Dukhan.

I have never seen an Arabic calligrapher in action, though I have seen a Chinese one. I visited dozens of bookshops, as is my wont, in the main cities of China and Japan, but I must say I was less interested in those big, perfectly ordered stores, from which I was driven by characters I couldn't understand, than in another kind of space and style that attracted me with their magnetic Oriental power. I was surprised to discover in Libro Books in Tokyo that Haruki Murakami had published several volumes of cybernetic correspondence with his fans. In Shanghai's Bookmall I liked leafing through the Chinese translation of *Don Quixote*. However, I particularly sought out a mixture of discovery and recognition in the Hutong tearooms on Philosophy Way, in a few gardens, in antique shops, in an old calligrapher's workshop. Perhaps it was because I could not understand them when they spoke to me, but I liked to hear the musical rhythms of Zhongyuan or Rui'an. Perhaps because I was denied any possible access to Japanese literature in the original language, I fell in love with the paper they used to wrap books, boxes of sweets, glasses or plates, and their extraordinary, refined art of paper-making.

I had another memorable tussle with the practice of haggling in an antiques shop in Beijing. After reviewing the dusty shelves packed with lovely objects, I was set on buying a teapot that seemed more affordable than the engravings, tapestries or vases. As we did not understand one another, the adolescent attending to me grabbed a toy calculator with huge letters and keyed in the price in dollars: 1,000. I snatched the device and keyed in my counter-offer: 5. He immediately went down to 300. I went up to 7. He asked for help from the owner, an impassive, ancient man with an alert gaze who sat down opposite me and, with a couple of whirls of his hands, indicated that it was now serious bargaining: 50. I went up to 10. He asked me for 40, 30, 20, 12. That was what I paid and I was so pleased with myself. He wrapped my teapot in white silk paper.

It was when I saw the American tourist in Budapest paying three times more for the same box that I understood the value of my own box and, above all, my would-be value as a tourist. It was in a Beijing market the day after my new purchase that I saw a hundred teapots identical to mine except that they gleamed, without a speck of dust, mass-produced, on a carpet, priced at one dollar, and realized that aura has to do with context (or was reminded of that yet again). Comparison and context are also factors when it comes to valuing the importance of a book, the text of which is tied to a specific moment of production. That is what literary criticism is doing continually: establishing comparative hierarchies within a specific cultural field. The framework of a bookshop is the physical place where we readers compare most. But to make that comparison we must understand the language in which the books we are looking at are written. And that is why, for me and so many other Western readers, the cultural ecosystems that we call the Orient, and the bookshops in which they are given material form, constitute a parallel universe that is at once fascinating and frustrating to navigate.

Paper was invented in China at the beginning of the second century AD. A eunuch, named Cai Lun was responsible: he made the pulp from rags, hemp, tree bark and fishing nets. Because paper was less exalted than bamboo and silk, it took centuries for it to establish itself as the best support for the written word and it wasn't until the sixth century that it travelled beyond the Chinese frontiers and until the twelfth that it reached Europe. In France its production coincided with the production of linen from flax fibre. By that time Chinese printers were using movable type, but the thousands of characters in the language stopped printing from really constituting a revolution, as would occur with Gutenberg four hundred years later. Nonetheless, as Martyn Lyons has noted in *Books: A Living History*, China had produced more books than the rest of the world put together by the end of the fifteenth century. Each volume: an object. A body. Matter. Secretions and silk-worm paper. Gutenberg had to perfect oil-based indelible ink by experimenting with soot, varnish and egg-whites. Forging type with alloys made from lead, antimony, tin and copper. In the following centuries a different combination was reached: nutshells, resin, linseed and turpentine. Though industrial production of paper was later standardized through the use of pine or eucalyptus wood, together with hemp or cotton, its manufacture from cotton rags, pure cellulose free of any bark, was still synonymous with quality in the eyes of the experts. Books depended on the rag-and-bone man until the eighteenth century, after which modern systems were developed to extract paper from wood pulp and the price of books was halved. Rags were cheap, but the process was expensive. In his studies on Baudelaire—as we have seen—Benjamin highlights the figure of the rag-and-bone man as a collector, as the archivist of everything the city has reduced to bits and pieces, flotsam from the city's shipwreck. As well as the analogy between fabric and the syntax of writing, between the rags used and the ageing of what is published, the closing of the circle is important: recycling, the reabsorption of rubbish by industry, so the information machine doesn't grind to a halt.

In the Orient, the idea endured for centuries that the best way to absorb the contents of a book was by copying it manually: that intellect and memory work with words in the same manner as ink does with paper.

VII
America (I): "Coast to Coast"

The classic route for the coast-to-coast ride begins in New York and ends in California. As this is a classic essay, a bastard child of Montaigne, this chapter will tease out the route, despite scant intermediate stops; a route that will inevitably develop into a journey as textual as it is audio-visual—though it is anchored quite firmly in particular bookshops that are *exemplary* in their way—through myths of American culture, a culture that is surely characterized, above all, by its creation of contemporary myths.

Nevertheless, most of them are individual and generally linked to significant spaces that often have collective connotations. Elvis Presley is a unique body in movement, and hence an itinerary, a biography; but he *is* also Graceland and Las Vegas. And Michael Jackson expressed himself spatially in Neverland, as Walt Disney did before him in his first theme park in California. Similarly, one can visit the cultural history of the United States in the twentieth century by focusing chronologically on certain emblematic places, *examples* of a complete picture one could never encompass. The 1920s saw the famous lunches at the restaurant in New York's Algonquin Hotel, where writers, critics and publishers like John Peter Tooley, Robert Sherwood, Dorothy Parker, Edmund Wilson or Harold Ross argued about aesthetics and the national and international publishing industry; in the 1930s the Gotham Book Mart established itself in the same city, specializing in the dissemination of experimental writers, organizing all manner of lectures and literary parties and gradually becoming a rendez-vous for avant-garde artists exiled from Europe; during the 1940s Peggy Guggenheim's New York Art of This Century gallery was the decisive launch pad for abstract expressionism as *the* form adopted by the nation's avant-garde; in the 1950s, the City Lights Bookshop in San Francisco brought onto the market some of the most representative books of that period and promoted them with launches and readings; under the leadership of Andy Warhol, The Factory in Manhattan stood out in the 1960s as a film studio, art workshop and home to druggy parties, and in the 1970s and early 1980s the nightclub Studio 54 picked up the baton.

Obviously, these are key places for their times. Especially on the East Coast, although one cannot understand the culture of the United States without the perpetual Coast-to-Coast movement: "I love Los Angeles. I love Hollywood. They're so beautiful. Everybody's plastic, but I love plastic. I want to be plastic," said Andy Warhol. If I had to choose a single building to symbolize, if only tangentially, intellectual life in the United States in the twentieth century, it would be the Chelsea Hotel, established in 1885 and still going strong. The list of its celebrities and important moments could begin with Mark Twain and end with Madonna (photographs of room 822 appear in *Sex*), not forgetting a few survivors of the Titanic, Frida Kahlo and Diego Rivera, Dylan Thomas' suicide in 1953, the writing of *2001: A Space Odyssey* by Arthur C. Clarke, the composing of *Blonde on Blonde* by Bob Dylan, the performing of "Chelsea Hotel" by Leonard Cohen and some scenes in *9 1/2 Weeks*. The hotel is like a bookshop. It is equally central to the history of ideas, as a meeting place for migrants, as a site for intense, solitary reading—which Edward Hopper portrayed so well—for writing and creation and the interchange of experiences, contacts and fluids. It is also at a crossroads between uniqueness and cloning, independence and chain, with a museum-like vocation. And it falls outside the institutional circuit and hence has a history hewn from discontinuities. Although more than a hundred and thirty years in New York guarantees the possibility of a chronologically structured narrative, as it has been visited—bombarded—by the biographies of hundreds of artists, the Chelsea Hotel and the other hotels where hundreds of artists have lodged on their endless travels, can only be recounted through a constellation of stories and dates.

The beat generation had to experience their fetish in the flesh, the Beat Hotel in Paris, that city which, in Burroughs' words, "is a disgusting hole for anyone without a dime," full of French people, "genuine pigs," but where he managed to finish *The Naked Lunch* and work on his cut-ups thanks to the facilities provided by a Frenchwoman, Madame Rachou, who ran that hotel without a name (9 rue Gît-le-Cœur) where he stayed with Ginsberg, Corso and other friends. When the movement transformed into the beat trend, the beat fashion, all things beatnik, that Paris hotel was christened the Beat Hotel. The same city that had watched the birth of Cubism half a century earlier via the brushes of Juan Gris, Georges Braque and Pablo Picasso now welcomed the postmodern éclosion of cut-ups and literary montage. After Tangier and Paris they continued to take drugs and create in the Chelsea Hotel in New York. Burroughs wrote that it was a place that "seemed to have specialized in the deaths of famous writers." The shooting of *Chelsea Girls*, Warhol's experimental film, can be seen as another turning point: the end of a particular way of understanding Romanticism, a wild, vagabond style, the start of serial production and the spectacular showcasing of contemporary art.

Were the beats good bookshop customers? They weren't, if one believes the legend. It is much easier to imagine them borrowing or stealing books, taking them for a while from the shelves of Shakespeare and Company rather than buying them. Indeed Whitman's bookshop—to judge by the copious correspondence—was above all a source of income: "The bookseller here, who is a friend of Ferlinghetti's, has fifty copies of my book in the window and sells several every week." The big book thief was Gregory Corso, who often tried to sell the books he had stolen the previous night the next morning. They were no doubt keener for second-hand than for new. And on reading originals, addicted as they were not only to chemical substances but also to the epistolary art, automatic writing, lyrical rhapsody and jazz rhythms. However, legends exist to be disproved: in Paris, for example, they took advantage of their access to Olympia Press books to acquire works by banned French and American authors. "Ferlinghetti sent me $100 yesterday, so we ate, I paid Gregory's 20 dollar back rent he's moved in with us temporarily," Ginsberg writes to Kerouac in a letter dated 1957. "We bought Genet and Apollinaire dirty book and a paper of junk and a matchbox of bad kief and a huge quart expensive bottle of perpetual maggi seasoning-soy sauce. While they lived in the Chelsea Hotel they went to New York bookshops like the Phoenix that mimeographed copies of Ed Sanders' magazine *Fuck You* and was behind a poetry collection in the form of chapbooks that included titles by Auden, Snyder, Ginsberg and Corso. Sanders himself opened the Peace Eye Bookstore in 1964 in an old kosher butcher's. It sold books as well as articles for counterculture fetishists, like a collection framed by the pubic hair of sixteen innovative poets or Ginsberg's beard. It quickly became a site for political activism and defended the legalization of marijuana amongst other things. On January 2, 1966 the police raided the shop and arrested its owner, accusing him of stocking obscene literature and lewd prints. Although he won his case, they never returned the confiscated material and he was eventually forced to close the bookshop.

If the abstract expressionists became heirs to the European avant-garde in the 1950s through a complex cultural, economic and political operation driven by institutions as different as the Museum of Modern Art and the CIA, it was thanks to a confluence of new sociological forces, new ways of understanding life and travel, music and art, as performative as the brushstrokes of Jackson Pollock, that the beat generation were to become the heirs to the lost generation and the French surrealists; namely, the usual suspects on the rue de l'Odéon. Until the Second World War, Gotham Book Mart was the United States equivalent of the original Shakespeare and Company. As we read in Anaïs Nin's diary, Frances Steloff's bookshop played the same role as Sylvia Beach's in Paris. The same infectious enthusiasm, the same support for nonconformist poetics: the shop lent Nin one hundred dollars, offered her all the publicity possible so she could self-publish *Winter* of *Artifice* and celebrate it with a launch party. But immediately after Hiroshima, Frances Steloff couldn't accept, or refused to acknowledge, the power of the beats and her renowned bookshop stayed anchored in the pre-war literary world. Art was a different matter: she found Duchamp an artisan to make the prototype of his famous suit-case-museum, and her shop window displayed an installation by him on the occasion of the launch of a book by André Breton and together with Peggy Guggenheim he designed another with the Art of This Century in mind. But the gesture that most emphatically defines the bookshop was the founding, in 1947, of the James Joyce Society, the first member of which was T. S. Eliot. Almost a decade later, when the Irish writer was still alive, Steloff dedicated an ironic shop-window display to *Finnegans Wake*, in the form of a wake in sync with the mood of the present. However, linking the bookshop to a dead author now afforded it a dangerously premature museum status, even if it was still a relatively young establishment (it opened in 1920 and did not disappear until 2007) when its Alma Mater was under fifty and destined to be a hundred-year-old bookseller.

One only has to read *In Touch with Genius: Memoirs of a New York Bookseller* to realize that, although the Gotham Book Mart always defended small reviews and fanzines, it also supported young writers and high-quality literature. The memoir remains faithful to its own roots and champions a particular kind of literature from the first half of the last century, the roll-call of which was defined by the publication of the anthology *We Moderns: 1920–1940*. The memoirs were published in 1975 and are reminiscent of Beach's: it was no coincidence that both booksellers were born in 1887 and devoted their lives to promoting the same authors, with James Joyce leading the way. The bookseller emulates her predecessor and assumes an observer's role ("I never approached my customers unless they looked as if they needed help") and is a collector of distinguished visitors. She met Beach in Paris and they worked together on several occasions. She ends on this note: "Our bookshops were often thought of as similar projects, but I never enjoyed the advantages she had."

In the 1920s and 1930s, the Gotham Book Mart became the focus for spotlighting books banned in the United States, an island where treasures by Anaïs Nin, D. H. Lawrence and Henry Miller could be found. This was the literary horizon that sealed its reputation and concentrated all its energy in terms of promotion. We find allusions to it in the private correspondence of these writers. For example, in a letter from the *Tropic of Cancer* author to Lawrence Durrell:

> Naturally the sales weren't very high, neither for *The Black Book* nor for *Max*. But they are selling slowly all the time. I myself have bought out of my own pocket a number of your books, which friends asked for. And now that the ban is off them, in America, we may get somewhere—through the Gotham Book Mart at least. In the next ten days or so I ought to have some interesting news from them, as I have written to them about the state of affairs. Cairns may not have had time to see you, his boat left the day after he

arrived. But he has a high opinion of you and all of us—a staunch fellow, full of integrity, somewhat naive, but on the right side. I count him a good friend and perhaps my best critic in America.

Gotham Book Mart and its famous slogan "Wise men fish here" appear in the graphic memoir *Are You My Mother?*, by Alison Bechdel, who writes "This bookshop has been here for ever, it's an institution." Culture has always circulated as much through alternative networks as established market channels and writers have always been the biggest shareholders in these parallel poetics. Nonetheless, it is worth underlining Miller's reference to Huntington Cairns to explore the complex relationships that exist between art and political power in the United States, given that the latter was both an excellent reader and a lawyer who advised the Treasury on the matter of importing publications that might be considered pornographic. In other words, he was a censor. Probably the most important one of his time. The letter, dated March 1939, ends on this rather startling note: "I'm a Zen right here in Paris, and I've never felt better or more lucid, secure and focused. Only a war could distract me from this." In another letter from that time to Steloff, he offers her the most recent first editions of *Tropic of Cancer* and *Black Spring*, and elaborates the idea even further: "My decision isn't based on fear of a war. I don't think there will be one this year, nor do I think there will be next year." Just as well he devoted himself to reading and not futurology.

In 1959, Gay Talese reported on the case of *Lady Chatterley's Lover*, a novel banned from the country until that year. A federal judge diluted the definition of obscenity that the Supreme Court had formulated two years before in the case of Samuel Roth v. the United States of America for dealing in pornography:

The liberation of the novel had actually been initiated by the courtroom efforts of a New York publisher, Grove Press, which had filed and won its case against the US Post Office, which until then had assumed broad authority in banning "dirty" books and other objectionable materials from being mailed in America. The courtroom triumph of Grove Press was immediately celebrated by advocates of literary freedom as a national victory against censorship and an affirmation of the First Amendment.

That was how one more of the infinite chapters on censorship in the history of culture was closed, as if we hadn't already had access to Diderot's words from the eighteenth century in his renowned *Letter on the Book Trade* (1763), a systematic dissection of how the publishing system works, from royalties to the writer's relationship with his printer, publisher and bookseller that, *toutes proportions gardées*, can be applied to a good number of areas into which the book trade is still divided legally and conceptually. Diderot, the driving force behind *L'Encyclopédie*, was himself forced to sell his library in order to provide his daughter's dowry. He was the author of other famous letters, like those to Sophie Volland or *Letter on the Deaf and Dumb*, and was possibly a lover of the Empress of Russia. After his death, one of the great novels of the modern era, his *Jacques the Fatalist,* was published. He wrote the following on the circulation of banned books:

> Please name one of those dangerous works that were banned, then clandestinely printed either abroad or in the kingdom that did not within four months become as available as any book which had been granted the *privilège*. What book is more contrary to good morals, to religion, to conventional ideas of philosophy and administration, in a word, to all

vulgar prejudices, and, consequently, more dangerous, than *The Persian Fables*? Is there anything worse? Yet there are a hundred editions of *The Persian Letters* and any student can find a copy for twelve sous on the banks of the Seine. Who doesn't own a translation of Juvenal or Petronius? There are countless reprints of Boccaccio's *Decameron* or La Fontaine's *Fables*. Is it perhaps beyond French typographers to print at the foot of the front page "By Merkus, in Amsterdam" like the Dutch printers? Multiple editions of *The Social Contract* are on sale for a crown by the entrance to the sovereign's palace. What does this mean? Essentially that we have always managed to secure these works; we have paid abroad the cost of labour that a more indulgent magistrate with better policies could have spared us rather than abandoning us to the black marketeers, who, taking advantage of our double curiosity, tripled by prohibition, have expensively sold on to us the real or imaginary danger which they exposed themselves in order to satisfy that curiosity.

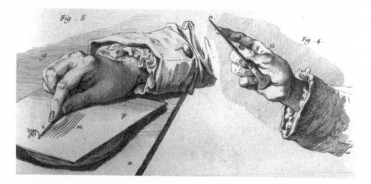

While small, and often short-lived, bookshops nourish the literary fiction that is outside the mainstream, bookshops that pride themselves on their huge size remind us that the publishing industry is not based on sophisticated books for a minority, but on mass production, just like the food industry. The New York bookshop equivalent to the Chelsea Hotel, in terms of independence, long life and symbolic importance, is possibly Strand, with its "eighteen miles of books," founded in 1927 by Benjamin Bass, who left it to his son Fred, who in turn bequeathed it to Nancy, his daughter, who handed the business on, in 2006, to her own children, William Peter and Ava Rose Wyden. The expression "family business" must have been coined with them in mind. Four generations and two premises: the original in "Book Row" on 4th Street, where in the good old days there were up to forty-eight bookshops, of which Strand is the only survivor, the current one being at 12th and Broadway. José Donoso wrote eloquently about its importance in an article entitled "A New York Obsession":

> I don't go to the big bookstores: I inevitably head to the Strand Bookstore on Broadway, corner of Twelfth Street, that cathedral of second-hand books where it is possible to find or order everything, and where on Saturday afternoon or Sunday morning one can see celebrities from the world of literature, theatre or cinema, in jeans and without make-up, looking for something with which to feed their obsessions.

I am interested in the insistence on the word "everything": the idea that there are bookshops like the Library of Babel, as opposed to those like Jakob Mendel's table in Café Gluck. Strand boasts that it houses one-and-a-half million titles. The record size, quantity and extent are advertising tools that suit a large number of bookshops in the United States, a naturally megalomaniac country. Its twenty kilometres of shelves were immortalized, in a manner of speaking, in the movie *Short Circuit*, by the memory machine, the reader on automatic pilot Number 5, when its voracious need for data provokes mayhem in the bookshop. On this side of the frontier, if we are to believe the hype, the biggest academic bookshop is in Chicago. In the months that I lived in Hyde Park I was a fan of the Seminary Co-op Bookstore on 57th Street, which was the best refuge when it was snowing, and was close to the university library. Its main distinguishing feature was "The Front Table," a coloured leaflet that reviewed the new main titles, though there was also a selection of other free cultural publications. It is one of those eminently subterranean bookshops in whose rooms one can spend long periods browsing in complete solitude. However, the store's main base was not on 57th Street but in the place where the cooperative was founded, in the basement of the theology seminary in the middle of the campus that is now home to the Becker Friedman Institute for Research in Economics. The University of Chicago has good reason to be proud of the twenty-four Nobel Prizes in economics its professors, guest researchers and ex-alumni have won, but nobody could give me many clues as to the movements of Saul Bellow and J. M. Coetzee through its corridors and neo-Gothic lecture theatres. On the other hand, in the digital magazine *Gapers Block*, I did find a bookseller, Jack Cella, who remembered that Saul Bellow used to love leafing

through newly arrived books, as they were being unpacked: the latest members of the community.

Conversely, Prairie Lights did find a way to benefit from the most famous creative writing programme in the country at the University of Iowa. Its web page lists the seven Nobel Prize in Literature winners who have visited them: Seamus Heaney, Czesław Miłosz, Derek Walcott, Saul Bellow, Toni Morrison, Orhan Pamuk and J. M. Coetzee. It was the personal project of Jim Harris, a graduate in journalism who decided to invest a small inheritance to open the bookshop in 1978. The present premises, now managed by his former employees, coincidentally occupy the space, that in the 1930s, housed a literary society where Carl Sandburg, Robert Frost and Sherwood Anderson used to meet. One of the former students of the renowned Writers' Workshop, Abraham Verghese, writes in the chapter about Prairie Lights in *My Bookstore: Writers Celebrate Their Favorite Places to Browse, Read and Shop*, that his booksellers were also teachers in a way: shaping sensibilities and treating him as a serious writer even when he had little confidence in his own potential. In the same volume, Chuck Palahniuk, focusing on Powell's City of Books, waxes ironic about the circuits for launching the latest books: Mark Twain died of stress on a reading tour.

The next stop on our coast-to-coast bookshop ride could be Tattered Cover in Denver, since that is where all important authors on tour in the United States stop, including Barack Obama. This project has been led since 1973 by the activist Joyce Meskis, a genuine civil-rights leader, who is so appreciated by her neighbours and customers that two hundred of them have helped the bookshop in its various moves, transporting boxes of books to other premises. Meskis applies a small profit margin, from one to five per cent on the price of the book, in order to be able to compete with the bookshop chains and show a customer that he or she is the great protagonist, the one who profits most. Her pleasant approach, moreover, is not restricted to personal and economic fields, but is also translated into dozens of armchairs that, according to the owner, are an attempt to remind the visitor that he is in a place like his own front room. Tattered Cover has always been characterized by its defence of civil rights, but in 2000 that struggle became national news when, by appealing to the aforementioned First Amendment, it succeeded in persuading the Supreme Court of Colorado to decide in its favour after the police tried to force Meskis to inform on a customer who bought a manual on how to manufacture methamphetamines. In the end it turned out to be a handbook on Japanese calligraphy.

Two thousand kilometres further on—leaving Las Vegas and Reno to our left—we come to another North American bookshop that no writer on tour can afford to miss, Powell's in Portland, whose appeal to Palahniuk must be that it is so like a vast brothel or a casino: countless interconnected rooms, a labyrinth in which each of nine rooms has its own name (Golden, Pink, Purple), like the characters in *Reservoir Dogs*. As in the Strand or other megalomaniac bookshops, quality is a treasure to be sought amongst layer after layer of quantity. No less than a million-and-a-half books. Visiting this shop is a journey you undertake with a map of the establishment as a guide, whether your aim is to find the Rare Book Room, with its eighteenth- and nineteenth-century volumes, or simply to go to the café to take a breather. Because Powell's in Portland is so famous owing to its size (it may *really* be the biggest in the world) it has become a tourist attraction and, as such, is constantly explored by visitors from the whole of this enormous country.

California lies to the south. To reach Los Angeles, where Quentin Tarantino's first film was shot and where so many fictional bookshops have been filmed and even built, you still have to cross Berkeley and San Francisco. It is really worth one's time to visit Moe's Books in the small university city, a building with 200,000 new, second-hand and old books; a bookshop with over half a century of history. It was founded by Moe Moskowitz in 1959, and consequently established itself as a cultural project in the political years of the 1960s with their protests against the Vietnam War. In 1968 the bookseller was arrested for selling shocking material (like Robert Crumb's comics and Valerie Solanas' books). After his demise in 1997, his daughter Doris took the helm, which she now shares with Eli, her own son, a third generation of independent booksellers. And in neighbouring San Francisco four important Californian bookshops await us: the oldest in the state (Books Inc.), the country's most famous (City Lights), perhaps the most fascinating one I know (Green Apple Books) and one of the most interesting that I have visited in terms of art and community (Dog Eared Books).

The history of the first goes back to the height of the Gold Rush in the middle of the nineteenth century, when, in 1853, the Swabian traveller Anton Roman started selling books and musical instruments to miners in Shasta City at the Shasta Book Store opposite the El Dorado Hotel. Its horizons were soon broadened by the addition of Roman's Picture Gallery: it was, after all, a desert where everything had to be created—culture, history, music, the frontier imagination. Four years later it moved to San Francisco, where the trade in texts and images was extended further to include its own printing and publishing business. It has changed place and name so often since then that all that remains is a slogan: "The oldest independent bookshop in the West." We have already mentioned the establishment is still run by Lawrence Ferlinghetti, and that was in terms of the French connection we find time and again in the history of

North American culture. Naturally, it is based in the centre of the city, next to Chinatown, Little Italy and most of the tourist icons. Green Apple Books, on the other hand, is off the beaten track, on the main roadway through the hybrid district of Clement. It appears in the novel *The Royal Family*, by William T. Vollmann, as it really is: the place to go in search of answers. The character in the novel opens the Buddhist Scriptures and reads: "Things do not come and do not go, neither do they appear and disappear; therefore, one does not get things or lose things." However, on my first visit to San Francisco, I went on a devout pilgrimage to City Lights because I still believed in my invisible passport; when I returned ten years later and they took me to Clement Street, I felt I was gaining something I would never lose.

Green Apple Books unequivocally demonstrates its vocation as a traditional neighbourhood bookshop in the balance it strikes between new and second-hand books, in its calculated improvization, in its dozens of passageways, uneven surfaces, connecting doorways and stairwells, in the dozens of handwritten reviews to guide readers and customers in their imminent choices, and its wooden floor. A bookshop is defined above all by what stands out: the posters, the photographs, the books recommended or displayed to draw attention. In Green Apple Books they have framed the "Open Letter" by Hunter S. Thompson, who came to San Francisco in the mid-1960s attracted by the magnetic power of the hippie movement. The stairs are dominated by a huge map of the United States, but there is also a section in the entrance called "Read the World," where new translations are recommended and displayed. And the right-hand-side wall in the basement is a genuine museum of African and Asian masks, the work of Richard Savoy, who opened the business in 1967, when he was barely twenty-five, his only work experience being as a radio engineer with American Airlines. But above all, it is about reading. In the labyrinthine bookshop on Clement Street you find them crouching down, almost hidden, as if confined within the cells of a Buddhist monastery or in the catacombs of the early Christians, of all ages and conditions, standing, squatting or sitting down: readers. And that is priceless.

A bookshop is a community of believers. This idea is nowhere better illustrated than in Dog Eared Books, which since 1992 has created a real atmosphere of empathy with the inhabitants of the Mission District. As well as magazines, books and graphic art, we find in the window on that street corner the perfect expression of the bond of love and respect a bookstore must create with its reader customers: an altar to the dead updated weekly by artist Verónica de Jesús. Anonymous neighbours, personal friends, writers and pop stars come together here. Famous readers or complete unknowns united by death and paid homage in a bookshop that, above all, feels itself to be part of a neighbourhood.

Somebody had stuck the photograph of Marilyn Monroe reading *Ulysses* on a bookshelf in Green Apple Books. The Hollywood Body reading the Mind of an Irish Writer exiled in Trieste or Paris. The United States reading Europe. In the old film *Funny Face* that kind of opposition had experienced an interesting twist. Under instructions from the editor of the magazine where he is employed, a fashion photographer played by Fred Astaire has to find models who harness beauty and thought, who "think as well as they look." The bookshop Embryo Concepts in Greenwich Village—invented in a Hollywood studio—is the place where the hunt-and-capture operation is carried out: there Fred Astaire meets Jo Stockton, a beautiful amateur philosopher (with the face and complexion of Audrey Hepburn), and persuades her to accompany him to a fashion show in Paris. She accepts, not because she is attracted by possible photo shoots, but because it might enable her to attend the classes of a philosopher who is an expert on "empathicalism." The inversion of traditional roles is striking in a film from 1957: he represents superficiality and she intellectual depth. However, in the end, as is only right in a musical, they kiss and the kiss erases, or at least freezes, all previous

142

friction. In *Notting Hill* there is an opposite starting point: he (Hugh Grant) runs an independent bookshop specializing in travel and she (Julia Roberts) is a Hollywood actress. While she browses upon entering his shop for the first time (the fictional Travel Book Company is in reality a shoe shop now called Notting Hill), he catches a book thief and politely explains the options: buying or returning the book hidden in his trousers. The thief recognizes the famous actress and asks her for an autograph; the bookseller, on the other hand, simply falls in love with her.

As an erotic space, every bookshop is the supreme meeting place: for booksellers and books, for readers and booksellers, for readers on the hoof. The familiar features shared by bookshops throughout the world, their nature as refuges or bubbles, means that encounters are more likely there than elsewhere. The strange sensation of knowing by the title that that book, published in Arabic or Japanese, is by Tolstoy or Lorca, or else by the author's photo or some kind of intuition. That shared experience of a re-encounter in some bookshop in the world. It is scarcely surprising that falling in love in a bookshop is a well-established literary and cinematic theme. In *Before Sunset*, the sequel to *Before Sunrise*, the story of the nine delicious hours the two protagonists shared in Vienna nine years earlier, while they were both travelling through Europe by train, they meet up again in Shakespeare and Company. True serendipity: he has become a writer and that is the place where American writers launch their books in Paris. The moment when he recognizes her has all the magic of a classic erotic performance. While he tells his audience the plot of a story he would like to write, a book made from a minimum present and maximum memories, which would endure for as long as a pop song, by way of flashbacks we enter the coded story of what that other superficial story would in fact retell, fragments of the preceding film, of that night in Vienna. Then he turns to his right and notices her. He recognizes her immediately. He becomes extremely agitated. They only have a few hours to pick up the thread they dropped almost a decade before. What prevails is a romantic attitude to the idea of the bookshop: it is a symbol of communication, of friendship, of love, as one detects in other products of popular culture, from novels like *The Shadow of the Wind* and *Seaglass Summer* to the romantic comedies *Remember Me* and *Julie & Julia*, both with scenes shot in Strand, and above all, *You've Got Mail*, where an independent bookshop is threatened by the branch of a chain that has opened around the corner while, simultaneously, the person running the first (Meg Ryan) and the manager of the second (Tom Hanks) are engaged in an epistolary relationship, though they don't know each other's names or faces.

Platonic love: love of knowledge. In an episode from the first season of the television series *The West Wing*, we are shown the police operation that is necessary whenever President Bartlett feels like buying the antiquarian books that are his passion. The majority of volumes he collects are from the nineteenth century or the beginning of the twentieth and are quite eccentric: on bear hunting, skiing in the Alps, Phaedrus and Lucretius. In contemporary fiction a bookshop signifies a space for the kind of knowledge that can't be found in official institutions—the library or university—because being a private business it avoids issues of regulation and because booksellers are even freakier than librarians or university lecturers. The expert who hoards banned or esoteric knowledge of fantasy or horror genres is now an alternative to the antiquarian shop with a secret room or basement. Several twenty-first century comics have as their refrain the idea of the bookshop as a clandestine archive, for instance *The Boys* by Garth Ennis and Darick Robertson, where the basement of a comics shop protects the memory of the super-heroic world, or *Neonomicon*, by Alan Moore, in whose bookshop you can buy all kinds of magic and sadomasochistic titles. This passage from the story "The Battle that Ended the Century," by H.P. Lovecraft, illustrates perfectly this idea of an alternative subculture on the periphery of the system:

> Mr Talcum's report on the event, illustrated by the well-known artist Klarkash-Ton (who esoterically depicted the fighters as boneless fungi) was reprinted after repeated rejections by the discriminating editor of the Windy City Grab-bag—as a broadside by W. Peter Chef. This, through the efforts of Odis Adelbert Kline, was finally placed on sale in the bookshop of Smearum and Weep, three and a half copies finally being disposed of thanks to the alluring catalogue description supplied by Samuelus Philanthropus, Esq.

However, it is not only occultism, magic, religion or books banned by the Inquisition or dictatorships that are to be found in bookshop alcoves and basements, any title bearing the aura of what is secret, little known, of a book for the happy few, the immense minority, connoisseurs, initiates, can be lodged in that crypt for relics or the strong box. When published, most books are democratically available to everyone: the price is calculated according to present-day factors. As the years go by, according to the good fortune of a work (and author), its rarity or aura, its status as a classic and its power as a myth, prices can rocket and enter an aristocratic dimension, or plummet until it is worth the same as any rubbish or cast-off. A book can be hunted down as much for its magical powers as its market value, and both factors often go together. When George Steiner, for example, reminisces about his discovery of the work of Borges, he does so in these terms:

> I recall an early connoisseur in the cavernous rear of a bookstore in Lisbon showing me—this was in the early 1950s—Borges' translation of *Orlando* by Virginia Woolf, his prologue to a Buenos Aires edition of Kafka's *Metamorphosis*, his key essay on the artificial language devised by Bishop John Wilkins, published in *La Nación* in 1942, and (rarest of rare items) *Dimensions of My Hope*, a collection of short essays published in 1926 but, by Borges' wish, never reprinted. These slim objects were displayed to me with an air of fastidious condescension. And rightly. I had arrived late at the secret place.

In Paris, the Alain Brieux bookshop combines antique books and prints with human skulls and nineteenth-century surgical equipment. An authentic cabinet of curiosities. The image of the antiquarian bookshop as a store of rarities oscillates between real referents and what is imagined, like all images with regard to that human impulse we call fiction. The Flourish and Blotts bookshop, in Diagon Alley, with a secret doorway behind Charing Cross Road in London, is one of the establishments where Harry Potter and the other student magicians go to stock up on school books at the start of each year. The Livraria Lello & Irmão in Oporto was used as the location for the filming of the screen version. On the other hand, Monsieur Labisse's bookshop in *Hugo*, which has similar charms, was made expressly for the film. Forty thousand books were required to that end. Alfred Hitchcock also used a Hollywood studio to recreate a bookshop, the Argonaut in San Francisco, in order to shoot a famous scene in *Vertigo*. The place is renamed the Argosy in the script and portrayed in the terms we have been outlining: an emphasis on its antiquity, a twilight scene, a supply of old volumes that preserve esoteric knowledge, and above all a specialist focus on the California of the pioneers that justifies the visit of Scottie, in his search for data on "sad Charlotte" as defined by Pop Leibel, the fictional bookseller inspired by the real Robert D. Haines, who befriended Hitchcock as a result of the latter's visits to the Argonaut. "She died," continues Leibel. "How?" enquires Scottie. "By her own hand," replies the bookseller, and he smiles sadly. "There are so many stories . . ." The screenplay reads: "It has gone dark inside the bookshop and the characters are reduced to silhouettes."

I have just discovered on the web that the Book City bookshop in Hollywood has closed down. It was a huge store of second-hand and bargain-basement volumes, a sort of West Coast replica of Strand, a stone's throw from the Boulevard of the Stars. They also sold screenplays. There were big cardboard boxes full of them, at $10, $5 or $1: for the price of pulp fiction, typed scripts, stapled together, scripts that were never filmed, perhaps never even read, bought by weight from the production companies that received them in excess, with black-and-white, opaque and transparent plastic covers, bound with plastic spirals, the same plastic that Andy Warhol so adored.

VIII
America (II): From North to South

T he Leonardo da Vinci Bookshop in Rio de Janeiro must be the most poeticized in the world. Márcio Catunda dedicated a poem, "A Livraria," to it, in which he describes the passageway leading to its entrails in the basement of the Edifício Marquês de Herval, the shop's windows luridly lit to create artificial daylight. The manager, Milena Piraccini, photocopied it for me, and I remember talking to her about the history of an institution that the previous year—it was the end of 2003—had existed for fifty years. We were standing next to two desks, where two huge calculators posed as fake cash registers—computers being banned—next to a complete collection of La Pléiade. Her mother, Vanna Piraccini, an Italian with a Romanian father, officially took over the business in 1965, after the death of her husband, Andrei Duchade, though she had managed it from the very beginning. Vanna faced the greatest adversities in the history of the trade and overcame them: economic recessions, the long military dictatorship, and the fire that completely destroyed the shop in 1973. Her friend Carlos Drummond de Andrade wrote: "The subterranean shop/exhibits its treasures/as if defending them/from sudden famines."

Right opposite, another bookshop has existed in that underground gallery, one that was also to become a landmark: Berinjela. Founded by Daniel Chomski in 1994—as I was told by the publisher Aníbal Cristobo, who lived in Rio at the beginning of the century, "It's a bookshop that reminds me of the one in the film *Smoke*: a meeting place for writers that can as easily lead to a recording label as a publishing house (it brought out the four issues of *Modo de usar*, perhaps the best contemporary poetry magazine in Brazil), or a quasi-clandestine den for the organization of championships of that mysterious game, *fute-botão*, or simulated football." I imagine a synergy is created between the two shops similar to the one once experienced on rue de l'Odéon. Though underground. Not anymore. I now discover while updating this book, which is eternally behind the times, that Leonardo da Vinci closed its doors in 2015. Half of that energetic embrace disappeared.

Also dedicated to the Livraria Leonardo da Vinci is a poem by Antonio Cicero, which I have a photocopy of and will translate:

Rio seemed infinite
to the adolescent I used to be.
Boarding the Castelo bus alone,
jumping off at the end of the line,
walking fearlessly,
to the centre of the forbidden city,
in a crowd that didn't notice that
I didn't belong there, and suddenly,
anonymous amid the anonymous,
feeling euphoric, sure I belonged there,
and them to me, going into side streets,
alleys, avenues, arcades,
cinemas, bookshops: Leonardo
da Vinci Larga Rex Central Colombo
Marreca Íris Meio-Dia Cosmos
Alfândega Cruzeiro Carioca
Marrocos Passos Civilizacão
Cavé Saara São José Rosário
Passeio Público Ouvidor Padrão

Vitória Lavradio Cinelândia:
places I didn't know before
opening onto infinite streets
corners forever spreading
across every city that exists.

An adolescent gazing at the city, its spaces and culture. An eroti-
cized, all-consuming gaze. For Juan García Madero, poetry—in
the beginning—is to be found in the arts faculty of the UNAM
and his room in the Lindavista suburb, but it soon shifts to
certain bars and cafés and visceral-realist haunts and bookshops
where he can satisfy his hunger on those lonely days when he
has nobody to talk to. In the opening pages of Roberto Bolaño's
The Savage Detectives, literature is sexualized: it could not be any
different given his adolescent protagonists. Juan discovers a
poem by Efrén Rebolledo, recites it, imagines a waitress riding
him and masturbates several times. Soon after, one of the literary gatherings ends in a blowjob. While drink and sex lord it
over literature by night, by day it is framed by bookshops, in the
labyrinth of which he tries to find "two disappeared friends":

> Since I don't have anything to do, I've decided to go looking
> for Belano and Ulises Lima in the bookstores of Mexico City.
> I've discovered the antiquarian bookstores Plinio el Joven, on
> Venustiano Carranza. The Lizardi bookstore, on Donceles.
> The antiquarian bookstores Rebeca Nodier, at Mesones and
> Pino Suárez. At Plinio el Joven the only shop assistant is a
> little old man who, after waiting obsequiously on a "scholar

from the Colegio de México," soon fell asleep in a chair next to a stack of books, supremely ignoring me. I stole an anthology of Marco Manilio's *Astronómica*, with a prologue by Alfonso Reyes, and *Diary of an Unknown Writer* by a Japanese author set during the Second World War. At Lizardi I thought I saw Monsiváis. I tried to sidle up next to him to see what book he was looking at, but when I reached him, Monsiváis turned and stared straight at me, with a hint of a smile, I think, and keeping a firm grip on his book and hiding the title, he went to talk to one of the assistants. Provoked, I filched a little book by an Arab poet called Omar Ibn al-Farid, published by the university, and an anthology of young American poets put out by City Lights. By the time I left, Monsiváis was gone.

The passage comes from a sequence (December 8, 9, 10 and 11 in the first part, "Mexicans Lost in Mexico (1975)") devoted to Mexico City's bookshops. And to bibliokleptomania: a practice as old as books themselves. There are descriptions of visits to Rebeca Nodier, Sótano, Mexicana, Horacio, Orozco, Milton, El Mundo and La Batalla del Ebro bookshops, the owner of the latter being "a little old Spaniard by the name of Crispín Zamora," to whom he confesses that "he stole books because he didn't have any money." In total: two books Don Crispín gives him, and twenty-four books he steals in three days. One of them is by Lezama Lima: we never find out the title. It is inevitable that in a novel about growing up bookshops are linked to voracious desire. In *Paradiso*, one of the characters suffers a sexual dysfunction related to books and a friend plays a joke on him, in a bookshop, in fact:

When the bookseller came in, he asked him, "And has James Joyce's *Goethe* arrived, the one that's just been published in

Geneva?" The bookseller winked at him, detecting the mocking nature of his question. "No, not yet, though we're expecting it any day now." "When it arrives, keep a copy for me," said the person talking to Foción, who did not get the joke referring to a book that had never been written. The voice was thick, coated with crispy meringue saliva, his sweaty hands and forehead revealing to boot the violence of his neuro-vegetative crises. "The same collection has a Chinese Sartre from the fourth century BC," said Foción. "Ask the bookseller to keep a copy of that too." "A Chinese Sartre who must have discovered a point of contact between *wu wei* and the nothingness of the Sartrean existentialists."

The crazy conversation about invented books continues, until the bookseller's interlocutor leaves the premises, walks up calle Obispo and goes to the hotel room where he is living. The narrator then tells us that he was suffering from "a sexual crisis that showed itself in an artificial, precipitate cultural anxiety that became pathological when he confronted the latest books in bookshops and the publication of rare books." Foción knows that and enjoys the passing lunacy in "the labyrinth," which is

what he calls bookshops. Erection. Fetish. The accumulation of stocks. The accumulation of erotic experiences is like a summation of different readings: their trace is virtual, pure memory. Stealing or buying books or receiving them as presents means possessing them: for a systematic reader, the shape of his library can be read, if not as a correlative of his whole life, at least as a parallel to his development as an individual during his youth, when that ownership is decisive.

Guillermo Quijas was eighteen when his grandfather, the teacher and bookseller Ventura López, asked him to take the manuscript of a book to a designer, then to the printer, and finally to collect the copies. As if those invisible bytes magically gave rise to volumes with pages, a smell and weight. However, that book did not come out of nothing: its existence formed part of a chain of meaning that went back as far as the 1930s, when a very youthful Ventura López worked his socks off to get a grant and graduate as a rural teacher and, some time later, as a primary-school teacher. He was sacked from his teaching job because he was the driving force behind an agricultural cooperative and had joined the Communist Party in 1949. Then, with the help of some comrades in a similar situation, he created a common fund that allowed them to open a book-cum-stationery shop, which became a cultural centre and literacy project that in the end also published a list of books about local culture. The Maestro died in 2002 but La Proveedora Escolar (the School Supplier) still exists in Oaxaca, thanks to his grandson's vocation. The two premises he inherited and five new ones coexist with Quijas' personal project, the Almadía publishing house, an Arabic word meaning "boat."

While orderliness tends to predominate in bookshops that sell new books, chaos reigns in second-hand shops: the disorderly accumulation of knowledge. The names of the bookshops themselves often suggest as much. In calle Donceles and adjacent streets we find Inframundo, El Laberinto or El Callejón de los Milagros (Miracle Alley), non-computerized shops where finding a book depends exclusively on a precarious system of classification, good fortune, inertia and, above all, the memory and intuition of the bookseller. Echoes of the grotto or cavern, of the Zarathustra bookshop described by Valle-Inclán—that universal Spanish-Mexican writer and exceptional brain—in *Bohemian Lights*: "Shelves of books kick up a fuss and cover the walls. Four horrific prints from a serial novel paper over the four panes of glass in the door. The cat, parrot, dog and bookseller converse in the cavern." In Caracas La Gran Pulpería del Libro (the Big Book Grocery Store) takes the reality of a subterranean, overflowing bookshop to the extreme: books pile up on the floor as if they had spilt from the shelves that had been attempting to contain them for years. When its owner, the historian and journalist Rafael Ramón Castellanos, who founded the business in 1976 and has ever since combined his work as a bookseller and writer, was asked in an interview how the books were classified, he replied that all attempts at computerizing them had failed and that everything was in "his memory and the memories of the shop assistants and his son Rómulo."

Having lunch one day with Ulises Milla in mid-2012 in a restaurant in Caracas, it suddenly occurred to me that this was the nearest I would ever be (at least phonetically) to Ulises Lima. The history he recounted was a history of exile from Spain and Latin America, a history of the migrations that populated that territory and built a culture, the route map to which Bolaño drew with jagged edges. Bookshops that transform deep, natural and overwhelming sorrow into individual memories that at once are human, brief and evanescent. Benito, Leonardo and Ulises: three generations of publishers and booksellers with a surname that suggests speed, distance and translation. *Ulises Milla*—I thought in that restaurant whose specialty is meat accompanied by cream cheese and avocado—is almost a tautology. He spent fifteen years dedicated to graphic design as a strategy to dodge the family inheritance. But he designed books. And he ended up as a publisher and bookseller.

Benito Milla was born in Villena, Alicante, in 1918, and, as the secretary of the Libertarian Youth of Catalonia, became part of the Republican exile of 1939. After a few years in Paris, where Leonardo, his first son, was born, his wife persuaded him ("my grandmother was behind all my grandfather's house moves") to move to Montevideo, where for sixteen years of his life, between 1951 and 1967, years of economic crisis and political conflict in Uruguay that would lead to the military dictatorship in the following decade, he moved from a book stall in the Plaza de la Libertad to founding Editorial Alfa and managing several cultural magazines.

My grand-father left Montevideo in 1967 to go to Caracas and take charge of the newly established Monte Ávila Editores," he told me. "Alfa continued in Montevideo in the hands of my father and in 1973 we moved to Buenos Aires taking the publishing house with us, from where we had to flee when the military came to power after the death of Perón; it wasn't until 1977 that Leonardo landed in Caracas and the Venezuelan period of Alfa began, which for administrative reasons had to be called Alfadil." His grandfather's project would have a third phase in Barcelona ("my grandmother is Catalan"), from 1980 until his death in 1987, when he was a partner in the Laia publishing house, which ended badly. Closing the circle. As if circles, which are tangible spaces, could be closed in the multiple time of parallel universes. He was the publisher of Juan Carlos Onetti, Eduardo Galeano, Mario Benedetti and Cristina Peri Rossi (here there is pride in Ulises' voice). He progressed from anarchism to a humanism whose key word—as Fernando

156

Aínsa has reminded us—was *bridge*: between human beings and their reading, between the countries of Latin America, between both shores of the Atlantic. And between different generations of the same family: Leonardo Milla, who as a child did not eat breakfast until the first book of the day was sold, transformed Alfa Publishing into the Alfa Publishing Group in the 1980s and expanded its network of bookshops (though he was never aware that it could be called a chain) with two premises known as Ludens and three as Alexandria

332 BC (the year when Alexander the Great defeated the Persians in Egypt and started building the city and its myth).

In 1942, while he was inventing his own shorthand-based language, which he called "*la taqui*," and in which several pieces of writing have been preserved that have yet to be decoded, Felisberto Hernández and his wife, the painter Amalia Nieto, opened a bookshop, El Burrito Blanco (the Little White Donkey), in the garage belonging to the house of his in-laws. Naturally, it was a failure. Montevideo is a mysterious city and the capital of a mysterious country full of such anecdotes and stories. It is similar to Switzerland and Portugal in its dimensions and pace of life. During the time I spent in Argentina, I used to travel to its neighbouring country every three months to renew my tourist visa, receive the payment for my articles in *El País* and visit its bookshops, full of decatalogued Argentinian and Uruguayan books you could only buy in Uruguay, like those published by the local branch of Alfaguara and by Trilce. With each fresh incursion I would uncover layers of a history of periodic migrations. And for that reason I wasn't surprised to find other traces in Peru, years later, on my only visit to its capital.

El Virrey (the Viceroy) in Lima has a corner with a chessboard between two armchairs. The ceiling fans turn slowly, like a gentle whisk. Everything is wood, books and wood, and the distant memory of exile. I decided to investigate the bookshop's history and asked the bookseller whether a summary existed somewhere. Her name was Malena. She said she would have to speak to her mother and gave me the email of Chachi Sanseviero, to whom I wrote straight away hoping for an interview. It wasn't possible, her excuse was that she had lost her voice, but with her reply she sent the text she had written for *Cuadernos Hispanoamericanos*. The Virrey in Lima opened its doors in 1973, financed by savings made in anticipation of a long exile from Uruguay. Its logo carries an image of the Inca Atahualpa holding a book in one hand and a *quipu* in the other: the two cultures' means of communication, the imposed and the original, united in a symbol of assimilation. Apparently, the Inca leader, when he found out that Chachi's book was sold as the genuine history of the true god, threw it to the ground in order to reaffirm that the truth was on his side of the pantheon. Chachi writes of the bookseller as somebody who always defers reading and transforms books into "eternal possibilities," "because, except in a very few cases, she never finishes reading them." She leafs through them, takes them to the counter, perhaps even home, to her desk or bedside table, where she will also not finish reading them.

The family tradition took another turn—or performed an unexpected pirouette, in 2012, when Walter and Malena opened their own bookshop, Sur, with the idea of following the path their father Eduardo had pioneered. I can see on the web that it is a delightful bookshop in which the straight lines of the shelves and the curves of the tables full of new titles join up to ensure that books are one hundred percent the protagonists. I sometimes think the Internet is the limbo where the bookshops I could not experience personally await me. A limbo of virtual spectres.

After spending the crucial years of his adolescence in Mexico City following the opposite route to the one followed by Ernesto Guevara twenty years earlier, in 1973 Bolaño travelled overland to Chile, where he intended to support Salvador Allende's democratic revolution. He was arrested a few days after the Pinochet coup and saved from likely death thanks to the fact that one of the policemen guarding him had been a schoolfriend. He returned to Mexico, also overland, to complete the life experiences that would nourish his first masterpiece. He died three months before I arrived in the Chilean capital. In El Fondo de Cultura Económica bookshop I bought the Planeta editions of *The Skating Rink* and *Nazi Literature in America*. In the latter, the most extended biographies belong to "the fantastic Schiaffino brothers" (Italo and Argentino, *alias* El Grasa) and Carlos Ramírez Hoffman (*alias* El Infame). Two Argentinians and a Chilean.

Although he spent most of his life in Mexico and Spain, which were the settings for much of his work, his literary allegiance was with the Southern Cone. As a Latin-Americanist he read work from the whole continent, as an adoptive Catalan and Spaniard he read his contemporaries, as a passionate admirer of French poetry he learned from its great masters, as a compulsive reader he devoured every title of world literature that was put before him. As a young man in Mexico he fought the figure of Octavio Paz regarding what he meant in terms of cultural politics; in his adult life he would periodically encounter enemies, literary translations from the armies against which he competed in his meetings in Blanes with fans of war games and strategy, but most of all he felt part of the tradition of the Southern Cone—if such a tradition really exists—and in his ambitious writer's mind that tradition was split in two: poetry and narrative. Chile and Argentina. Bolaño the poet felt close to Lihn and Nicanor Parra. And near and far at the same time, as regards Pablo Neruda, who is to Chilean poetry what Borges is to Argentinian narrative: they are Monsters, Fathers, Saturns devouring their children. It is strange that Juan Rulfo was not held in esteem by Mexican writers in the second half of the last century whereas Paz certainly did play that portentous, castrating role (as did Carlos Fuentes). I often wonder what would have happened if Rulfo had become the principal model for Spanish-American writers at our *fin de siècle* and had occupied the place history reserved for Borges. The rural, anachronistic, minimalist Rulfo, who looked to the past, who believed in History, who said no, in the place of urban, modern, precise Jorge Luis Borges, who looked to the future, who scorned history, who said yes. In "Dance Card," Bolaño tells the story of his copy of *Twenty Love Poems and a Song of Despair* and the long distance it travelled between towns in southern Chile, and then around Spain, and recounts how, at the age of eighteen, he had read the great poets of Latin America and how his friends were split into supporters of Vellejo or Neruda and that he was thoroughly isolated as a supporter of Parra. In this account, Chilean poetry is organized into dancing partners, with descendants and disciples of Neruda, Huidobro, Mistral and De Rokha, and the heirs to Parra and Lihn. His alliance with Parra and Lihn is fissured by the crevice through which the hugeness of Neruda slips, an influence no poet in the Spanish language can escape. In "Dance Card," recognition of Neruda's political inconsistency leads to a crazy excursus on Hitler, Stalin and Neruda himself and a genuinely Bolañesque passage on institutional repression and common graves, the International Brigades and the torture racks. Ultimately, Neruda remains a contradictory mystery.

When his sister gave him Neruda's book, Bolaño was reading the complete works of Manuel Puig. In terms of story-writing, it was in "Sensini" (from *Telephone Calls*) where he best defined his connections with committed left-wing Argentinian literature, through the character of Antonio Di Benedetto. The theoretical essay "The Vagaries of the Literature of Doom" was where the Chilean positioned himself in respect to Argentinian literature and tackled the question of the canon with no holds barred. Bolaño repeatedly recalled his debt to Borges and Cortázar, without whom one cannot grasp the encyclopedic ambition of his work, his interest in auto-fiction and the short story or structure—the paths opened up by *Hopscotch*, *The Savage Detectives* and *2666*. It was in the latter in particular that he put himself forward as Borges' greatest heir, launching into severe criticism of his Argentinian contemporaries and the short cuts and roundabout routes they took to elude the centrality of Borges: those who followed Osvaldo Soriano, those who saw

Roberto Arlt as the Anti-Borges, those who championed Osvaldo Lamborghini. That is to say, many writers who aren't mentioned, like Ricardo Piglia and César Aira.

During three of four days I spent in Santiago, I decided, no doubt in a rush, that Libros Prólogo was the bookshop that most interested me. I noted at the time:

> It's not as big as the University Library in Alameda (with its wall-to-wall carpets and 1970s look) or the Chilean Feria del Libro chain, nor does it have the charm of the second-hand bookshops on calle San Diego, but it is well stocked and in calle Merced, next to a cinema, theatre and café and close to the antiquarian and second-hand bookshops in calle Lastarria.

I haven't kept any other notes. I remember it as a place of resistance, a centre that nourished cultural life during the dictatorship, but I have no way of proving that or finding out. Nothing on search engines. Perhaps mine was the delirium of a traveller seduced by *By Night in Chile*, the novel in which Bolaño constructs the lunatic discourse of the priest Sebastián Urrutia Lacroix, who under the pseudonym Ibacache celebrates the savage, reactionary poetry of Ramírez Hoffman in *Distant Star*, and who in the final part of the novel remembers the lessons in political theory he gave to the Junta and the literary conversations that took place in the house of Mariana Callejas. The character is inspired by the Opus Dei priest José Miguel Ibáñez Langlois, who wrote for the newspaper *El Mercurio* under the pseudonym Ignacio Valente, and was the author of books of philosophical and theological theory (*Marxism: Critical Vision*, *The Social Doctrine of the Church*), of literary criticism (*Rilke, Pound, Neruda: Three Key Contemporary Poets*, *Reading Parra*, and *Josemaría Escrivá, as a Writer*) and of poetry with a fondness for oxymoronic titles (*Dogmatic Poems*). He was not merely the

most important literary critic during the dictatorship and the transition, he also gave seminars on Marxism to the Junta. Pinochet was one of his pupils. Pinochet: the reader, the writer, the lover of bookshops. Ricardo Cuadros has written:

> Ibáñez Langlois has never acknowledged or denied his presence at the literary soirées held by Mariana Callejas in her big mansion in the wealthy district of Santiago that she ran with Michael Townley, her husband, a D.I.N.A. agent; those get-togethers were real enough and in the mansion's basement among others Carmelo Soria, a Spaniard working for the U.N., was tortured to death.

That basement in a "taken over" house is the exact opposite of what the great majority of bookshops in the world have been, still are and will ever be. There were and are bookshops with the

163

name of Cortázar's story in several cities (Bogotá, Lima, Palma de Mallorca . . .) because the title has been freed of its associations with the story and come to mean "space taken over by books." The story, on the contrary, speaks about how they disappear. The narrator of "House Taken Over" regrets that no new books have arrived in the French bookshops in Buenos Aires since 1939 so he cannot continue to nourish his library. If the political interpretation of the story is correct and the writer is creating a metaphor for Peronism as the invader of private spaces, it is no coincidence that the first part of the house that is taken over is the library. The protagonist's sister is a weaver; he is a reader. But after the first takeover, reading is gradually erased from his life. When the house is definitively taken over and brother and sister close the door for good, they will only take with them the clothes they are wearing and a clock, but no books; the cord has been cut.

When I returned to Santiago de Chile ten years later, I felt in a state of trance, like a sleepwalker pursuing the threads he had left trailing on his daily treks, as if in some invisible intrigue. It was twelve noon, the blistering sun beating down, and I was walking through the Lastarria district on the verge of unconsciousness. By chance I had just found the hostel where I had lodged on my only other previous stay: perhaps it was the charge of erotic memories that had provoked my mechanical promenade, which suddenly cloaked my skin with that of someone else, the person I used to be in my early twenties. I was not surprised to find myself suddenly outside Libros Prólogo, the bookshop that had most caught my attention at that time, on those days that followed my nights in that hostel with its games, kisses and topsy-turvy sheets. Nor to see Walter Zúñiga behind the counter, as if he had been waiting for me in the same shirt, with the same wrinkles, for ten years.

"What are you reading so intently?" I asked after browsing for a few minutes.

"A biography of Fellini written by Tullo Kezich that I bought yesterday in La Feria," he replied, with the big ears of an old man who really knew how to listen. "It's odd, I've had this book here for ages, two copies, in fact, it's extraordinary that I've never sold it."

"So if you've got it, why did you buy another?"

"It was so cheap . . ."

We spoke for a while about the bookshop's other branch, which had closed down, and he confessed that the ones that really worked for him were his Karma bookshops, "specializing in fortune-telling, Tarot, New Age and martial arts." I asked him for a copy of a recently published book about a pioneering cybernetic project during the years of Salvador Allende's government—

"*Cybernetic Revolutionaries*," he interrupted as he keyed away. "Right now I'm out of stock, but I'll get you a copy in a couple of days." He'd already picked up the telephone and was ordering it from the distributor.

A few minutes later he bade farewell with a flourish, giving me his card. He had corrected the telephone number in black ink. It was exactly the same card I had in my bookshop archive. The same red-lettered typography. "Libros Prólogo. Literatura-Cine-Teatro." It was a very strong connection to the traveller I had been ten years earlier. Everything had changed in the city and myself except for that card. Touching it stirred me from my sleep-walking, dragged me violently from the past.

It was only natural then that I should walk fifty paces, cross the street and enter Metales Pesados and the wavelength of the present. Not an ounce of wood in the place, only aluminium shelves, the bookshop as a giant Meccano structure that welcomes books as fervently as an ironmonger's or a computer lab.

There was Sergio Parra, in black suit and white shirt, a dandy and all sinew, sitting on a folding metal chair behind a café terrace table. I asked him for a copy of *Leñador* (*Woodcutter*) by Mike Wilson, which I had been hunting for in the Southern Cone for months. He handed it to me, hiding his gaze behind his hipster glasses. I asked him about Pedro Lemebel's books and now he did finally look me in the eye.

I later discovered he had led the campaign backing Lemebel for the National Literature Prize, and that they were friends, poets and neighbours. But right then I had eyes only for the large poster of the writer and performer and the prominent display of all his books, because every bookseller deals in visibility. He recommended a couple of books I did not have, which I bought. "Metales Pesados is more of an airport than a bookshop. At any moment Mario Bellatin will walk in, or a person Mario or somebody else encouraged to drop by, friends of friends from all over the world, many leave their suitcases here, because they've already checked out of their hotels and still have a few hours to go up the hill or to the Fine Arts Museum. As I practically live here, since I work here from Monday to Sunday, I've become a kind of reference point."

The bookshop as an airport. As a place of transit: for passengers and books. A pure to-and-fro of readings. Lolita, on the other hand, far from the centre, in a corner of a residential district, preferred people to stay on. It also had a writer behind the counter: Francisco Mouat, whose passion for football had led him to devote a corner of the shop to sport. The journalist Juan Pablo Meneses accompanied me to that recently opened bookshop and showed me how Juan Villoro, Martín Caparrós, Leonardo Faccio and other friends we had in common were sparsley represented there by volumes they had devoted to the small round god. Mouat's gaze is gentle and his gestures are pleasant and welcoming, despite his intimidating height; I'm not surprised he has full reading groups every Monday, Wednesday and Friday.

"A book a week is quite a pace," I say.

"We've kept that up for some time. I have readers who've been following me for years. When I opened Lolita I brought them to this new place."

Loyalty is there in the slogan: "We can't live without books." Loyalty is on the logo: a dog that belonged to the Mouat family looks at us, embossed on the spines of the books from the Lolita publishing house.

When on my last day in Santiago I finally visited Ulises, that space so crammed with books and where the distorting mirrors that reflect the shelves and the volumes *ad infinitum* fly over and multiply you, wonderful ceiling mirrors designed by the architect Sebastian Gray, perhaps because I was in one of the most beautiful, most Borgesian of bookshops in the world, the fourth vertex of an invisible rectangle, I reflected on how the other three, Libros Prólogo, Metales Pesados and Lolita, gave material form to the three tenses of every bookshop: the archival past, the transitory present and the future of communities united by desire. How, in combination they form the perfect bookshop, the bookshop I'd take to a desert island.

And I suddenly remembered a scene I had forgotten. A repeated, distant scene, a fading echo or call from a black box in the depths of the ocean, and the accident. I must be nine or ten, it's a Friday evening or Saturday morning, my mother is in the butcher's or bakery or supermarket, and I'm killing time in the neighbourhood newsagent's, Rocafonda, on the far outskirts of Barcelona. As there is no bookshop in the district, I am a true addict of kiosks, with their superhero comics and video-game magazines, of the tobacconist's Ortega, who has quite a window display of his collection of books and popular educational magazines, and of this newsagent's in the same street where the

Vázquez brothers and other schoolfriends live. There are less than a hundred books at the back, behind the stands with coloured card, birthday cards and cut-outs; I have fallen in love with a manual for the perfect detective; I remember its blue cover, and I remember (the power of the memory upset me, as I left Ulises and got into a taxi and headed to the airport) how I would read a couple of pages every week, standing up, how to get fingerprints, how to make an identikit photo, every week until Christmas or Sant Jordi finally arrived and my parents gave me the book I had so coveted. At home, after I had read it, I realized I knew it by heart.

How could I ever forget that for a large part of my childhood I had two vocations: writer and private detective. Something of the second stuck with me in my obsession for collecting stories and bookshops. Who knows whether we writers are not, above all, detectives investigating ourselves, Roberto Bolaño characters?

On the terrace of the Café Zurich in Barcelona, Natu Poblet, who runs La Clásica y Moderna bookshop in Buenos Aires, told me that in 1981, when she gave up architecture to devote herself to the family business, with two years still left in the last

dictatorship, they organized classes in literature, theatre and politics on their premises, given by people banned from the university, like David Viñas, Abelardo Castillo, Juan José Sebreli, Liliana Heker, Enrique Pezzoni or Horacio Verbitsky. "The classes turned into literary conversations, my brother and I took along wine and whisky, lots of people came and the conversations went on till very late," she told me as she downed a glass of Jameson's; that was when the idea of harnessing a bookshop to a bar was born. It implied a one-hundred-and-eighty degree turn. Her grandfather, the Madrid bookseller Don Emilio Poblet, founded the Poblet Brothers chain in Argentina at the beginning of the twentieth century. Her father, Francisco, opened La Clásica y Moderna in 1938 with his wife, Rosa Ferreiro. Brother and sister Natu and Paco took charge of the business after their father died in 1980. That was the year the Junta ordered the burning of a million-and-a-half books published by the Publishing Centre of Latin America. After seven years of activity in the catacombs, with democracy re-established, they commissioned the architect Ricardo Plant to radically transform the space that has been a bar and a restaurant ever since, as well as a bookshop and exhibition and concert hall. ("The first three years we opened twenty-four hours a day, but then we started to have problems with night-time drunks and decided to adopt a more conventional timetable.") Since then, actors like José Sacristán and singers like Liza Minnelli have performed there. The piano was a present from Sandro, a habitué of La Clásica y Moderna from its frenetic heyday, and whose life story can be gleaned from the titles of some of his albums: *Beat Latino*, *Sandro de América*, *Sandro . . . Un ídolo*, *Clásico*, *Para mamá*.

"I often dream of Dad's bookshop," Natu Poblet confessed as she drained her glass and we began a long stroll round night-time Barcelona. In Río de Janeiro, Milena Piraccini talked to me straight away about the importance that Vanna, her mother, attached to personal contact with each of her customers, a character trait that could be explained by her forebears in Europe. In Caracas, Ulises Milla told me about his Uruguayan family and about other booksellers from Montevideo and Caracas like Alberto Conte, who had taught him so much. Chachi Sanseviero writes:

> My teacher was Eduardo Sanseviero, a great bookseller and disciple of Don Domingo Maestro, a notable Uruguayan bookseller. Eduardo's weakness was chess, history and antique books. But he also liked poetry and had the strange gift of bringing poems into the conversation like funny stories. An unrepentant communist, in times of despots, he enjoyed organizing small conspiratorial cells. But at the end of the day, he went back to his feather duster and arranging his books.

Bookselling is one of the most secretive traditions. It is often a family affair: Natu, Milena, Ulises, Rómulo and Guillermo and Malena, like so many other booksellers, are the children and even grandchildren of booksellers. Almost all of them began as apprentices in the bookshops of their parents or other traffickers in printed paper. Rafael Ramón Castellanos remembers that, when he reached Caracas from the interior of Venezuela, he worked in a bookshop, Viejo y Raro (Old and Rare), that belonged to a former Argentinian ambassador: "Later on, in 1962, I created my own bookshop with the knowledge that I had acquired," the Librería de Historia that preceded the Gran Pulpería de Libros.

Isn't the figure of the bookseller rather odd? Is it not easier to understand writer, printer, publisher, distributor, or even a literary agent? Might this oddness explain the lack of genealogies and anatomies? Hector Yánover, in *Memorias de un librero*, illuminated these paradoxes thusly:

This is the book of a bookseller with pretensions. These are the first lines of that book. These words constitute the first on the first page. And all these words, lines and pages will make up the book. Do you, hypothetical readers, have any idea how horrific it is for a bookseller to have to write a book? A bookseller is a man who reads when he rests, and what he reads is book catalogues; when he goes for a walk, he stops in front of the windows of other bookshops; when he goes to another city, another country, he visits booksellers and publishers. Then one day this man decided to write a book about his trade. A book inside another book that will go to join the others in the windows or on the shelves of bookshops. Another book to arrange, mark, clean, replace, remove definitively. A bookseller is the being who is most aware of the futility of a book, and of its importance. That is why he is a man torn apart; a book is a commodity to buy and sell and he now constitutes that commodity. He buys and sells himself.

Yánover ran the Librería Norte in Buenos Aires and, according to Poblet, was the city's most important bookseller in the final quarter of the last century. His daughter Débora now holds the reins of the business. He was also responsible for a renowned record collection which included recordings of Cortázar and Borges, amongst others, reciting their work. When the author of "The Pursuer" travelled back to his country, he made the Librería Norte his centre of operations: he spent the whole of his first day in the city there and it was there that his admirers could leave letters and parcels of books for him. I do not know if those records are in a corner of the Bolaño Archive, or whether he listened to their dead voices as he did to opera and jazz. However, the bookshop that marked the life of the author of *Ficciones* was the Librería de la Ciudad (the City's Bookshop), which was next to his house, on the opposite side of the calle Maipú, inside the arcade that goes by the name of Galería del Este. He visited it daily and gave dozens of free lectures there on matters that appealed to him. He launched in its rooms the titles of the Library of Babel, the collection that he was commissioned to edit by the Milanese publisher Franco Maria Ricci and which was partly co-published by the bookshop itself. Borges and Cortázar didn't meet in a bookshop, but in a private house on Diagonal Norte, where the younger of the two turned up to discover that the Maestro had so liked his story "House Taken Over" that it was already at the printer's. They met again, years later, in Paris, by which time they had both been honoured by the Académie Française. I have not been able to identify the bookshop where Cortázar bought *Opium*, by Jean Cocteau, the book that changed his work—I mean life, though I did find the interview with Hugo Guerrero Marthineitz in which the author of *Nicaraguan Sketches* tries to justify Borges' behaviour during the military dictatorship, which he had backed to restore order and during which he had also defined himself as "a harmless anarchist" and as "someone revolutionized" who was "against the state and against the frontiers of states" (as his biographer, Edwin Williamson, amongst others, has pointed out). And who chose to die in Geneva. Cortázar's rhetorical gymnastics are similar to those we find in Bolaño's lines on Neruda: "He wrote some of the best stories in the world history of literature: he also wrote *A Universal History of Infamy*."

That is the model for *Nazi Literature in America*, a book written at a distance from Europe. Complexity is the most difficult thing to judge: Ibáñez Langlois defended Neruda and Parra, both fathers of Bolaño the poet, and supported the career of Raúl Zurita, whose poems written in the sky seem to have influenced the work of the infamous Ramírez Hoffman to some extent. It is not too far-fetched to read the whole of Bolaño's oeuvre as an attempt to understand his own damaged, lost, re-formed library, with as many absentees as fellow travellers, compounded by the distance that did not allow him to fully understand what was happening in Chile while at the same time affording him the critical lucidity necessary for oblique readings; a complex, contradictory library, decimated by house moves and rebuilt in European bookshops. We read in one of the articles collected in *Between Parentheses*:

As for my father, I don't remember him ever giving me a book, although occasionally we would pass a bookshop and at my request he would buy me a magazine with a long article in it on the French electric poets. All those books, including the magazines, along with many other books, were lost during my travels and moves, or else I let people borrow them and never saw them again, or I sold them or gave them away.

But there's one book I'll never forget. Not only do I remember when and where I bought it, but also the time of day, the person waiting for me outside the bookshop, what I did that night, and the happiness (completely irrational) that I felt when I had it in my hands. It was the first book I bought in Europe and I still have it. It's Borges' *Obra poética*, published by Alianza/Emecé in 1972 and long out of print. I bought it in Madrid in 1977 and, although Borges' poetry wasn't unfamiliar to me, I started to read it that night and didn't stop until eight the next morning, as if there was nothing in the world worth reading except those poems, nothing else that could change the course of the wild life that I'd lived until then, nothing else that could lead me to reflect (because Borges' poetry possesses a natural intelligence and also bravery and despair—in other words, the only things that inspire reflection and keep poetry alive).

There is no ideological questioning. There is no moral suspicion. Borges quite simply does not belong to the revolutionary tradition, though that does not reduce his value. He is less problematic than Neruda. In *Advice from a Follower of Morrison to a Joyce Fanatic*, Bolaño and A.G. Porta refer insistently to the bookshops of Paris: as a bunker for political reading (the character reads *El Viejo Topo* [*The Old Mole*] there): the bookshop as an invitation to the moral voyeur ("I have always liked bookshop windows. The surprise you get looking through the glass and finding the latest book by the biggest bastard or the most out-and-out hoodlum"), the bookshop as something beautiful in itself ("I have been in two or three of the most beautiful bookshops I have ever seen"). One of them, although he does not say which, must be the fake Shakespeare and Company. Its remake. His idea of filming *Ulysses* in Super-8 comes from that visit.

In "Vagabond in France and Belgium," one of the stories in *Last Evenings on Earth*, a character by the name of B. takes a walk round the second-hand bookshops in Paris and in one on rue du Vieux-Colombier finds "an old copy of the magazine *Luna Park*" and the name of one of its contributors, Henri Lefebvre, "suddenly lights up like a match struck in a dark room." He buys the magazine and goes into the street to lose himself, as Lima and Belano did before him. Another name, this time of a magazine, now lights up on this page I am writing: *Berthe Trépat* was the name he and Bruno Montané chose for a mimeographed magazine they published in Barcelona in 1983. The light doesn't last long, but it is enough for us to be able to read about certain traditions of writers and booksellers, certain genetics common to the history of literature and bookshops, that is, culture, forever shifting—like a geological fault, like a quake—between the candle and the night, between the lighthouse and the night-time firmament, between the distant star and dark sorrow.

IX
Paris Without Its Myths

In 1997, the film director and writer Edgardo Cozarinsky premiered the docu-drama *Fantômes de Tanger* ("Phantoms from Tangier"), with dialogue in French and Arabic. The protagonist is a writer in crisis who reaches the shores of Africa in pursuit of some of the American spectres who have appeared in this book as well as the French kind, who helped create the myth of the international white city. Their opposite is a boy looking for a way to emigrate to Spain. Literary Tangier coexists with sameful poverty-stricken Tangier. Writing and sexual tourism interpenetrate in an uneasy relation where the boundaries are clear: customer and worker, exploiter and exploited, the one who has francs or dollars and the one aspiring to have them, with French as the *lingua franca* on both sides, in conflict despite an apparent dialogue. Traces of Foucault and Barthes fuse with those of Burroughs and Ginsberg, and converge in the brothels where young Moroccans have always prostituted themselves.

The documentary side of the film focuses on the survivors of a would-be golden era that is suddenly shown to be quite murky. "Everybody has passed this way," says Rachel Muyal in the Librairie des Colonnes, "through this bookshop." She follows this with an anecdote I expect she has told many a time: "I saw Genet, who was drinking coffee with Choukri, when a shoeshiner came and asked whether anyone wanted their shoes polished, then Juan Goytisolo took out a 500 franc note. That must have been a couple of years before he died." Three contemporary myths in a single frame that only the bookseller seems to want to preserve intact. "I feel no nostalgia whatsoever for International Tangier, it was a wretched period," says Choukri when interviewed in the film. Bowles badmouths Kerouac and the rest of the beats. Juan Goytisolo told me he never met up with Genet in Tangier. And Rachel Muyal insisted years later that I had got it wrong. Cozarinsky's film is in one of my trunks from my travels; it is a VHS copy you can't watch anymore.

Who knows who is right, if indeed anyone can be? All myths exist to be shattered.

I am particularly interested in the reading that the author of *For Bread Alone* gives of that gilded foreign legion. From his perspective, infected by his economic dependence on Paul Bowles, who helped him write his first book and translated it into English, thus launching him on the international market, Genet was little more than an impostor, not only because his poverty wasn't comparable to the real poverty suffered by the Moroccans Choukri describes so graphically in his autobiographical books, but also because he did not speak a word of Spanish, which meant his tales of the underbelly of Barcelona or certain parts of Tangier could not be taken seriously. He christened Bowles the Recluse of Tangier, because he spent the last years of his life lying in bed and because, in his view, he never properly connected with the Arab culture around him. One only has to read Bowles' letters to realize that though he physically resided in Morocco, his cultural focus was the United States, where he psychologically spent more and more time as he grew old. Nonetheless, his intellect allowed him to grasp that the traffic of Anglo-Saxon writers was turning the city into a masquerade, into a fiction, that the visitors never bothered to explore Moroccan society in depth, no doubt because he himself was not interested in such a total view. In a 1958 article titled "Worlds of Tangier" he wrote: "A town, like a person, almost ceases to have a face once you know it intimately," and that requires time. At some point in the next forty years he decided a certain level of intimacy was enough. In 1948, in a letter written in the Hôtel Ville de France in Tangier, Jane said to him: "I still like Tangier, maybe because I have the feeling of being on the edge of something I will some day be part of."

Paul Bowles in Tangier begins: "How ridiculous. I think nothing is more ridiculous than that exaggerated nostalgia for the Tangier of yesterday and that longing for its past as an international zone." However, I keep wondering why Choukri really wrote this book, or its twin, *Jean Genet and Tennessee Williams in Tangier*, and to what extent his desire to demystify is unconnected to the fact that he will only continue to be read in the West if he writes about French or Anglo-Saxon celebrities. It is not clear and never will be. The pain seeping through his words is undeniable; he is not killing his own father in vain: "He liked Morocco, not Moroccans." The Cabaret Voltaire edition of his portrait of Bowles was launched in Tangier, in mid-2012, by its translator, Rajae Boumediane El Metni, and by Juan Goytisolo in the Librairie des Colonnes, naturally.

In her notebook of reminiscences, Muyal describes her first encounter with Choukri ("We were having dinner on that wonderful scented terrace of La Parade restaurant on a summer's night with my pretty young cousins when a young stranger tried to give us flowers. When he saw we weren't going to accept them, the boy started to take the leaves off and eat the petals"), the way her reading of *For Bread Alone* shocked her because she was unaware such extreme, blatant poverty existed in her own city, and how he often intervened in the conversations on literature and politics in his many visits to her establishment. Tahar Ben Jelloun translated him into French, so he had two exceptional translators in the two most important languages in the world of publishing, but *For Bread Alone* soon became one of those books famously banned in its own language: "Two thousand copies were sold in a few weeks, I received from the Ministry of the Interior a note banning the sale of this book in any language." Nevertheless, fragments of his book were published in Arabic in newspapers in Lebanon and Iraq. When the narrator of Teju Cole's *Open City* (2011) asks another character to recommend a book "in keeping with his idea of authentic fiction" to him, the latter doesn't hesitate to jot down the title of Choukri's most famous book on a scrap of paper. He contrasts him with the more lyrical Ben Jelloun, "an Orientalist" integrated into Western circles, while Choukri "stayed in Morocco, lived with his people," never leaving "the street." In another novel published a year later and on another continent, *Street of Thieves*, Mathias Énard also has his narrator defend the Moroccan writer's magnetic qualities: "His Arabic was hard like the blows his father rained down on him, hard as hunger. A new language, a way of writing I thought was revolutionary." An American writer of Nigerian extraction, Cole hits the nail on the head when he defends the importance of Edward Said for our understanding of Oriental culture: "Difference is never accepted." What

Choukri did for the whole of his life was precisely that: he defended his right to be different, critically, approaching and distancing himself from those who gave him recognition, as always happens in life, in every kind of negotiation.

In *Never Any End to Paris*, Enrique Vila-Matas talks of Cozarinsky, whom he often came across in cinemas in the French capital: "I remember I admired him because he knew how to combine two cities, two artistic allegiances," he notes in fragment 65 in his book. He is referring to Buenos Aires, Cozarinsky's birthplace, and Paris, his adopted city; but the fact is that a tension between two places exists in all his work: between Tangier and Paris, between the West and the East, between Latin America and Europe. "I especially admired his book *Urban Voodoo*, an exile's book, a transnational book, employing a hybrid structure very innovative in those days." If Bolaño had a re-encounter with Borges in Madrid, Vila-Matas discovered the stories of Borges in the Librairie Espagnole in Paris, thus following in Cozarinsky's footsteps: "I was knocked out, especially by the idea—found in one of his stories—that perhaps the future did not exist."

I am also knocked out by the fact that this idea was suggested to him on premises run by Antonio Soriano, a Republican exile who nourished the hope of a future without Fascism. The Spanish diaspora sustained the cultural activity of resistance at the back of the Librairie Espagnole, as well as in Ruedo Ibérico. The project is linked, as is almost always the case in the history of a bookshop, with a previous one, the Librairie Espagnole León Sánchez Cuesta, established in 1927 in five square metres of the rue Gay-Lussac, with two window displays: one devoted to Juan Ramón Jiménez and the other to young poets like Salinas and Bergamín. It was run by Juan Vicéns de la Llave, who went so far as to consider publishing books from Paris in Spanish (the first was *Ulysses* in Dámaso Alonso's translation). In order to return to Madrid during the turbulence in Spain in 1934, he left the bookshop in the hands of a former employee, Georgette Rucar. But during the war, as the official responsible for Republican government propaganda in its Paris Embassy, he used the premises as a centre for spreading the ideas that were being crushed by Franco's army. After the Second World War, it was Rucar—as related by Ana Martínez Rus in "San León Librero: las empresas culturales de Sánchez Cuesta" ("St Leon the Bookseller: the Cultural Enterprises of Sánchez Cuesta")—who made contact with Soriano, who had settled down as a bookseller in Toulouse, to suggest he took over the stocks of the old bookshop. This book could be called *Metamorphoses* rather than *Bookshops*.

When Vila-Matas arrived in Marguerite Duras' attic in 1974, he witnessed the last gasps of that world, if not the photographs of its autopsy. As a mature man, the author of *A Brief History of Portable Literature* revises his initiatory experience in Paris, the Paris of his personal myths, as in Hemingway, Guy Debord, Duras or Raymond Roussel, where everything evokes a splendid past that has necessarily been lost, which paradoxically never goes out of fashion. Because each generation relives a kind of Paris in its youth, which can only be gradually demystified as one grows older.

Someone had drawn a graffiti sketch of Duras in the emergency exit of La Hune, with her famous saying on the left: *"Faire d'un mot le bel amant d'une phrase."* It took me five trips to Paris to discover that of its hundreds of bookshops, perhaps the best three are Compagnie, L'Écume des Pages and La Hune. On my previous visits, apart from persisting with Shakespeare and Company, I stepped inside all those I encountered, but for some reason these three never figured on my itineraries. So, before leaving on my last trip, I asked Vila-Matas himself for advice, and once back, I sought out and found them. I discovered the poster of Samuel Beckett (his arboreal face) against a cork background on a wall in Compagnie. And the art deco shelves in L'Écume des Pages. And that unlikely staircase and the whitest of columns in the middle of La Hune, part of a 1992 renovation, the work of Sylvain Dubuisson. The first is located between the Sorbonne, the Musée de Cluny and the Collège de France. The second and third are close to the Café de Flore and Les Deux Magots and open every day until midnight in Saint-Germain-des-Prés, perpetuating the old bohemian tradition of combining bookshop, wine and coffee. Although in the time separating the writing of these lines from their publication, La Hune has already disappeared (like the odd other bookshop mentioned here).

When Max Ernst (after marrying Peggy Guggenheim and becoming a habitué of the Gotham Book Mart), Henri Michaux (after bidding farewell to literature to concentrate on his painting), or André Breton (after his American exile) returned to a

Paris without the bookshops on rue de l'Odéon, they found in La Hune a new space where they could converse and browse. One need only reflect that, in 1949, the same year it moved to 170 Boulevard Saint-Germain, La Hune hosted the exhibition and auction of the books, manuscripts and furniture from Joyce's Paris flat (he died in 1941), together with part of Beach's archive on the publishing history of his masterpiece, on behalf of the writer's family. Soon after, Michaux began to experiment with mescaline and the graphic work he produced led, in the mid-1950s, to books like *Misérable miracle* and exhibitions like "Description d'un trouble," in the Librairie-Galerie La Hune. Its founder, Bernard Gheerbrant, who died in 2010, was a key figure in Paris' intellectual life and directed the Club des Libraires de France for over a decade. Because of his importance as a publisher of texts and graphic art, his archives are preserved in the Centre Pompidou, where he curated the exhibition "James Joyce and Paris" in 1975. After a brief, failed attempt to explore other approaches, La Hune finally shut down in June 2015. The artist Sophie Calle decided she would be the last person to buy a book. The last customer, the last reader. Her performance began a period of mourning that has yet to end, and which isn't restricted to the shop's Parisian customers but is felt by all of us who passed through its doors and left ever so slightly, but quite definitely, changed.

Like most of the bookshops mentioned in this essay, these three are fetishes in themselves as well as places for the exhibition of fetishes. A fetishism that goes beyond the classical Marxist definition, according to which all goods are phantom fetishes, which hide their status as manufactured commodities and maintain an illusion of autonomy in relation to their producers; a fetishism promoted by the agents of capital (publishers, distributors, booksellers, every one of us) who have fun (as we do) championing cultural production and consumption as if they were not subject to the tyranny of *interest*; a fetishism that borders on the religious and even on the sexual (in Freudian terms): the bookshop as the deconsecrated temple where idols, objects of worship are housed, like a store of erotic fetishes, of pleasure. The bookshop as a partially deconsecrated church transformed into a sex shop. Because a bookshop feeds on the energy generated by objects that seduce by virtue of their accumulation, by the difficulty of defining demand that becomes palpable when one finally locates the arousing object, demands an urgent purchase and a possible subsequent reading (the arousal doesn't always survive, but the percentages of the price of the book, the expenditure and profits do remain behind, like ashes).

Dean MacCannell has analyzed the structures of tourism and provided a basic schema: the relationship of the *tourist* with the *view* via the *marker*. Namely: the visitor, the attraction, and all that denotes it as such. The crucial element is the marker that indicates or creates the value, importance and interest of the place and transforms it into something potentially touristic. Into a fetish. The shop selling would-be antiques in Beijing was a fantastic marker. Although the value may be iconic in the first instance, in the end it becomes a discourse as well: the Eiffel Tower is first a postcard, a photograph, and then the life of its inventor, the history of its controversial construction, of other towers in the world, the topography of Paris that surrounds it and which can be seen from the top. The most meaningful bookshops in the world highlight, with greater or lesser degrees of subtlety, the markers that add commercial potential or transform them into tourist spots: antiquity (*founded in, the oldest bookshop in*), size (*the biggest bookshop in, so many miles of shelving, so many hundreds of thousands of books*) and the chapters in the history of literature to which they are linked (*the base for such and such a movement, visited by, the bookshop where X*

bought, visited by, founded by, as can be seen in the photograph, bookshop linked with).

Art and tourism are similar in their need for luminous signage that draws a reader towards the work. Michelangelo's David would attract little attention if it were an anonymous work in the municipal museum of Addis Ababa. After successfully publishing *The Golden Notebook* in 1962, Doris Lessing sent a new novel two years later to several publishers under the pseudonym of an unpublished writer, and all but one rejected it. In the case of literature, publishers first generate the markers, through the blurb on the back cover or the press release, and critics, the academy and bookshops soon create their own, which will determine the book's fortune. Sometimes authors themselves do this, consciously or unconsciously, by structuring a narrative around the conditions in which the work was produced, or the state of their lives at the time. Suicide, poverty or the context of the writing are elements often incorporated into the marker. That narrative, the legend, is one of the factors that allows a text to survive, to live on as a classic. The first part of *Don Quixote* was supposedly written in prison and the second part was a reaction against the usurping of Cervantes' characters by Avellaneda; the reading of *A Journal of the Plague Year* as if it were not a novel; the legal procedings against the authors of *Madame Bovary* and *Les Fleurs du mal*; the broadcast reading of *The War of the Worlds* and the collective panic created by the chronicle of that apocalypse; Kafka on his deathbed ordering Max Brod to burn his work; the manuscripts of Malcolm Lowry that were burnt, that disappeared; the scandal surrounding *Tropic of Cancer* and *Lolita* and *Howl* and *For Bread Alone*. The marker is sometimes unpredictable and created years later. That is the case of novels rejected by multiple publishers like *One Hundred Years of Solitude* or *The Conspiracy of Fools*. Of course, it was not used as a selling point when they were—finally—published, but when they were a success it was recovered as part of the mythical narrative: their *predestination*.

The stories behind several books published in Paris like *Ulysses*, *The Naked Lunch* or *Hopscotch*, have clearly been fetishized and now constitute commonplaces in the history of contemporary culture. For the beat generation, which felt it was heir to symbolism and the French avant-garde, *Ulysses* was the obvious reference point for their idea of a rupture. Written in the heady days in Tangier, shaped by Ginsberg and Kerouac, completed in France, *The Naked Lunch* was submitted to Maurice Girodias, the Olympia Press editor on the Left Bank, who did not understand such rubbish and declined to publish it. But eighteen months later, when the publication of a few fragments had begun to build the novel's reputation for outrageous obscenity, a marker, Girodias' interest in the manuscript was rekindled. By that time, the success of *Lolita* had made him a rich man, and Burroughs' novel, the writing of which was by now but a hazy memory for the author, helped him become even wealthier. He fitted within a fine French tradition: that of the dealer in scandalous books, often banned for being pornographic or obscene, which were published in Switzerland and entered France thanks to the inevitable bribes at the border, and which in the twentieth century were published in Paris and even reached the United States through a motley array of picaresque subterfuges.

Kerouac wrote about *On the Road*: "*Ulysses*, which was thought to be a difficult read, is today thought of as a classic and everybody understands it." We find the same idea in Cortázar, for whom this tradition is central and who linked himself to Paris, not only in the way he contextualized the first part of his novel, but also through a partial rewriting of Breton's *Nadja*. In a letter to his publisher, Francisco Porrúa, he cites the same reference point as a paradigm of difficulty, rupture, resistance and distinction for his contemporaries: "I reckon this must always happen; I'm not familiar with the contemporary reviews of *Ulysses*, but I expect they went something like this: 'Mr Joyce writes poorly, because he doesn't write in the language of the tribe.'" Like *The Naked Lunch*, *Hopscotch* has a structure that functions on the basis of fragments, collage, chance, and has a politically revolutionary intent: the destruction of the bourgeois ordering of discourse, the exploding of literary conventions, which are so similar to social conventions. That is why in his letters to his publisher the writer tries to spell out the marker, the discourse that should guide the reading of the book. We have to make an effort to imagine the difficulties of a publishing relationship carried out through epistolary means, the delays, misunderstandings, lost letters (for example, the envelope with the mock-up of the novel Cortázar put together that went missing):

> I would prefer it if this book was not highlighted as "a novel." That would rather cheat the reader. I know only too well that it is a novel and that its intrinsic value resides in its appearance as a novel. But I wrote it as an anti-novel and Morelli takes it upon himself to say that and spells it out very clearly in the passages I quoted to you previously. As a last resort, I think one should emphasize what we might call the axiological sides to the book: the persistent, exasperated denunciation of the inauthenticity of human lives [. . .] the irony, the derisory tone, the self leg-pulling whenever the writer or characters descend into philosophical "seriousness." After *On Heroes and Tombs*, you must understand that the least one can do for Argentina is to denounce at the top of one's voice the ontological "seriousness" of the creeps our writers aspire to be.

Hopscotch immediately struck a chord with his youthful contemporaries. The Paris he delineates revives the classic image of the bohemian city; the proliferation of topographical details transforms it into a possible guide for cultural tourists, which has been underlined by editions that incorporate a map or a list of the writer's favourite cafés; its encyclopedic dimension (literature, painting, cinema, music, philosophy . . .) means a reading can never be exhaustive. A classic work is one that always offers a new reading. A classic is a writer who never goes out of fashion. And Paris was precisely where fashion as we understand it today was born, so it is hardly surprising that, at least until the 1960s, and thanks to a continuous stream of artists from all over the world, it was able to sustain a seductive horizon of expectations for certain readers with respect to certain works, a fetishist aura. Pascale Casanova writes:

> Gertrude Stein neatly summed up the question of the localization of modernity in a single sentence: "Paris," she writes in *Paris France* (1940), "was where the twentieth century was." As site of the literary present and capital of modernity, Paris to some extent owed its position to the fact that it was where fashion—the outstanding expression of modernity—was created. In the famous *Paris Guide* of 1867, Victor Hugo insisted on the authority of the City of Light, not only in political and intellectual matters, but also in the domain of taste and elegance, which is to say of fashion and everything modern.

The logic that partially explains the relationship between Greek and Roman culture in antiquity, when revision, continuity, imitation, importing and usurping were the ways in which the empire could secure cultural hegemony, in which the original myths were reformulated (from Zeus to Jupiter) and the epic rewritten (from Homer's *Iliad* and *Odyssey* to Virgil's *Aeneid*) could be our model for understanding the relationship between the United States and France in the contemporary era. Although London is also of cultural importance in the nineteenth century, Paris establishes itself—as we have seen—as the international centre of literature and the visual arts. In the 1920s and 1930s, celebrities like Hemingway, Stein, Beach, Dos Passos, Bowles or Scott Fitzgerald found in Paris the feeling of being at the heart of bohemia. For a whole generation of American intellectuals—the names selected are a tiny fraction of all those who travelled to Paris and took their ideas from there like so many souvenirs—

France was a model for cultural *grandeur* and an adopted heritage. If Hemingway was right and the French capital was "a moveable feast," it is hardly surprising that he would leave in the 1930s, when the Nazis came to power in Germany and the Second World War finally broke out. Picasso stayed in Paris, where he created the marketing system for contemporary art; Beach also stayed on and Hemingway returned as a soldier of liberation. But the majority of the French avant-garde and American novelists met up again or for the first time in New York, together with artists, gallery owners, historians, journalists, architects, designers, film directors and booksellers. The same city where, after big exhibitions like the Van Gogh or Picasso, the Museum of Modern Art began to create from that subsoil its own narrative for contemporary art, first raising the standard of abstract expressionism and then pop art with Andy Warhol and The Factory leading the way. The 1950s and 1960s are fascinating because the American writers who are most in step with the times continue to visit Paris. But their approach is different. When Kerouac or Ginsberg travel to France they do so—reversing the route taken by Bowles—by stopping off at Tangier, as if one city was not more important than the other. Kerouac's mother tongue was French, Ferlinghetti translates surrealists like Jacques Prévert. Later on, other American writers with strong links to the fictionalized bookshop, like Paul Auster—Mallarmé's translator—will travel to Paris, but the key literary reference points for the later generations are American, not European. Paris has been transformed into a Library of Universal Literature, while San Francisco, Los Angeles, Chicago or New York are continually launching bookshops destined to be some of the most important cultural centres of the twentieth century. Whether for good or for evil, and not on United States soil, whether as ambassador or intruder, one such is the second Shakespeare and Company.

In the documentary "Portrait of a Bookstore as an Old Man," somebody says that George Whitman was the most American person he had ever known, because he was completely pragmatic and penny-pinching: the tasks in the bookshop had to be carried out by young lovers of literature, who did not receive a wage in return, but a bed, a meal and—though he does not say this—wonderful experiences working and living in Shakespeare and Company, in the heart of Paris. Whitman simply created the dream of every young American reader and the bookshop responded to a stereotype—like Flourish and Blotts in *Harry Potter*—and became a tourist attraction with a very powerful marker, as important for a student of literature as the Eiffel Tower or the Mona Lisa, with the extra bonus that you could *live* there; like the map in *Hopscotch*, it allowed you to create a literary space, to transform it into a home or hotel. "Living the dream"could have been its slogan. And it did so via a conceptual and commercial operation that went back to Sylvia Beach's original Shakespeare and Company, which can be seen from two perspectives: after-life or legacy on the one hand; appropriation, or even usurping, on the other. Whitman said in an interview: "She never found out anything about our intentions. We waited until she died, because if I'd have asked her and she'd said no, I couldn't have taken over the name after she died. All the same I think she would have said yes." Clearly, if he chose not to call his business Maison des Amis des Livres, it was because, being Anglo-Saxon, he saw the commercial potential of a name that guaranteed a stream of tourist pilgrims. Also because of his insecurity.

The film depicts an unstable, despotic bookseller as prone to handing out wounding insults as to being poetically maudlin, using his guests as volunteers in a labour camp whose working conditions he never properly spells out. Despite the bookshop's handsome income and the five-million-euro-estimated value of its building, he was a frugal, bohemian bookseller who spent no money on clothes or food, and had no social or emotional life outside his picturesque kingdom. We will never know whether he burnt his hair off with two candles in front of the camera because he was suffering from senile dementia or to save on the cost of a barber. And he called his daughter, who now owns the business, Sylvia Beach Whitman.

To be fair, his portrait should be balanced against the chronicle *Time Was Soft There: A Paris Sojourn at Shakespeare & Co.* by Jeremy Mercer. Whitman appears in its pages as an unstable old man, but also as very generous, affectionate and dreamy, ready to share his essential books and personal memories of Paris with anyone who sleeps in his bed. Memories of Lawrence Durrell drunk at night after spending all day writing *The Alexandria Quartet*; of Anaïs Nin, who may have been the bookseller's lover; of Henry Miller, the beat generation and Samuel Beckett, who naturally only ever paid silent visits; of all the books and magazines sent on their way by his bookshop; of Margaux Hemingway, whom he guided through Paris in search of her grandfather's city.

What was and is Shakespeare and Company? we wonder, after seeing the film and reading the book. A socialist utopia or a business run by a miser? A tourist icon or a really important bookshop? Was its owner a genius or a madman? I don't think answers exist to these questions and, if they do, they won't be black or white, but a range of greys. It is quite clear that L'Écume des Pages and La Hune are not mythical bookshops in the sense that Shakespeare and Company is, and are not internationally renowned, and that forces us to ask yet again: what is the stuff that myths are made of? And, more particularly, how can we demythologize them?

I too am guilty of contributing to this process of mythologizing (mystifying). All journeys and all readings are partial: when I finally visit Le Divan—the origins of which go back to the 1920s in Saint-Germain-des-Prés, which was resurrected by Gallimard in 1947 and lodged in the 15th Arrondissement since 1969—and research its history; when I discover Tschann, founded in Montparnasse in 1929 by two friends of the leading lights in the artistic life of that once-bohemian, now chic district, the Tschanns, whose daughter Marie Madeleine was a decisive supporter of Beckett's work in France, I will at last be able to repay the persistence of translator Xavier Nueno, who I hope will introduce me to the present person in charge, Fernando Barros, who emphasizes in interviews that he is equally conscious of the past and the future of the bookshop; when—finally—my reading or travels or friends take me to other neighbourhoods and new bookshops, my topography of Paris will change and, with it, my discourse. In the meantime, I accept the limitations of this impossible, future encyclopedia, which is as full of chiaroscuro as they all are and which I am perpetually writing.

X

Book Chains

From 1981, Shakespeare and Company also becomes a chain of independent bookshops, with four branches in New York and all near university campuses. Although many universities have their own bookshop that sells manuals, reference books and, above all, textbooks, T-shirts, tracksuits, mugs, posters, maps, postcards and other tourist items linked to the university experience, Barnes & Noble has colonized this market with more than 600 college bookshops in the United States, in addition to 700-plus branches in cities, each with its own Starbucks (it remains to be seen how this figure will be affected by the 2013 announcement that a third of their premises would be closed down over the next ten years). Although the first bookshop with that name opened in 1917, the Barnes family has had interests in the printing industry from the 1870s. A hundred years later, it became the first bookshop to advertise on television; and, in the present century, the main threat to the survival of small independent bookshops. Which is quite paradoxical, because many businesses that start with a single shop tend to multiply and become links within the same brand or chain. Many well-established chains also began as single, independent bookshops. Long before it owned dozens of branches throughout Mexico, Ghandi, opened in 1971 by Mauricio Achar, was a bookshop to the south of the capital. The biggest chains in Brazil originated as projects started by immigrants: Joaquim Inácio da Fonseca Saraiva, from the Trás-os-Montes region of Portugal, opened the first Saraiva in 1914, although at the time it was called Livraria Acadêmica; the first Nobel was founded in 1943 by the Italian Claudio Milano (in 1922 his grandson adopted a leasing system and branches quickly multiplied); the Livraria Cultura was the idea of a German-Jewish immigrant, Eva Herz, and arose from the idea of the book-lending service she started from the front room of her house in 1950; it did not become a bookshop until 1969. The three empires were born in the same city, São Paulo, and spread throughout the country. Family Christian Stores now have almost 300 branches. In 2012 they donated more than a million Bibles to be distributed by missionaries across the world, but the Zondervan brothers began with remainders from de-catalogued stock on a farm in the 1930s. Their growth came down to the success of their cheap editions of out-of-copyright religious works, including a number of English translations of the Bible.

Thanks to the fact that Holland was a haven for Calvinists and to the absence of religious and political censorship, it became one of the great world book centres in the sixteenth and seventeenth centuries. The Elzevir family was pre-eminent among its printers, and, between 1622 and 1652, published authentic pocket-book classics annotated by academics. Martyn Lyons reminds us that the 1636 edition of Virgil's complete works was such a success it had to be reprinted fifteen times. Pocket-sized classics were known as "Elzevir editions," regardless of whether the Elzevirs were the publishers. Despite their success, this kind of publication was aimed at the literate elite. One has to remember that the *Encyclopédie*, a genuine bestseller that sold almost 25,000 copies, was mainly purchased by the nobility and clergy, the social classes whose pillars it was undermining. Ordinary people read mostly slim chapbooks, booklets full of drawings, or the *bibliothèque bleue* bound with the blue paper from sugar packets distributed by itinerant sellers known as *colporteurs* in French, *Jahrmarksttrödlers* in German and *leggendaio* in Italy. The lives of saints, nonsense stories, farces, parodies, drinking and rabble-rousing songs, myths and legends, tales of chivalry, harvest calendars, horoscopes, gaming rules, recipe books and even abbreviated versions of universal classics were the real bestsellers before the explosion of the romantic and realist novel in the nineteenth century and its spread in the form of mass-produced serial fiction.

The book as a money-making success began with Walter Scott and was consolidated by Charles Dickens and William Thackeray. The volume of sales of books by Scott was so high in Europe that, starting in 1822, his novels appeared simultaneously in English and French. In 1824 a parody of his fictions, *Walladmor*, where Scott himself figured as a character, was published in Germany because, as we all know, there is no better guarantee of success than imitation or parody. The Lévy brothers launched a collection of works that cost one franc in Paris in the middle of the century. Michel and Calmann had become wealthy by commercialising opera libretti and plays and opened one of the great nineteenth-century bookshops on the Boulevard des Italiens, where there was a bargain section. Apart from investing in the bookshop, they also poured money into railways, insurance companies and public services in the colonies. In the same period, Baedeker and Murray popularized travel guides that could now be bought like so many other kinds of books from infinite outlets: grocery stores, kiosks, itinerant sellers, independent bookshops and chains. In *Reading and Riding*, Eileen S. DeMarco has studied the network of Hachette bookshops in French railway stations, a project that lasted almost a century, from early in 1826 to the outbreak of the First World War in 1914, with the launching of the first premises in Paris in 1853 en route. Trains rapidly became the principal vehicle for books: their trucks transported paper, printing presses, spare parts, the workforce, writers, finished books from one city to another and, above all, readers. The chain based its efficiency, for the first time in history, on the contracting of female shop assistants, *femmes bibliothécaires*, given that the initiative was called Bibliothèque des Chemins de Fer. In the letter Louis Hachette sent to the owners of the main railway companies in France to persuade them of the viability of his proposal, he emphasized its pedagogic nature, since the light, portable books would have an educational aspect as well as providing entertainment for the journey. By July 1853, forty-three branches had opened their doors and offered close to 500 titles. The following year they set up the daily press that, over time, would become their main source of income. And three years later they incorporated part of the output of other publishing companies, thus maintaining a monopoly on sales in stations. This was extended to the Métro network at the end of the century.

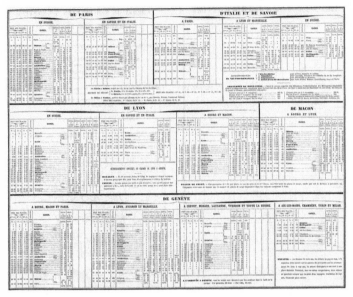

The A. H. Wheeler bookshop chain had a monopoly on book sales in stations in India until 2004. Like that of Hachette—which is now a transnational publishing group that shifts 250 million books a year—its railway history makes for fascinating reading. The first branch opened its doors in 1877 in the station of Allahabad, after Émile Moreau and his partner, T. K. Banerjee, borrowed the name from someone who probably never stepped on Asian soil: Arthur Henry Wheeler, who owned a bookshop chain in London. An agreement with the Indian government gave them a monopoly on the distribution of books and newspapers, with an evident social and educational intent: for over a century it was the principal way culture reached the most remote parts of the country, where A. H. Wheeler was often the only bookshop for many miles around. In 1937, with independence on the horizon, Moreau transferred his share in the ownership to his Indian friend and partner, whose family has run it ever since. The company entered the present century with some 600 sales points in almost 300 stations, but it lost the monopoly in 2004 in a nationalistic political move by Lalu Prasad Yaday, the Minister for Railways, against the British resonances in the Indian company's name. However, the decision was revoked six years later: the bookshop chain is too emblematic for it not to be treated as part of the country's cultural heritage.

As Shekhar Krishnan explains in an article in the *Indian Express* that is my source for this information, "See you in Wheeler's" is a common expression in Mumbai. The name is deeply rooted in the country's daily life. It is common to meet friends or acqaintances in its bookshops and kiosks while buying a newspaper before boarding the train and sharing the journey home. Conversations about politics and literature have for decades been organized around those sales points where people stand and drink their cups of tea.

Rudyard Kipling was born in Mumbai and his fate was linked to the name "Wheeler," which was shared by the editor of the *Civil and Military Gazette*, the first daily newspaper that the future writer worked for at the age of seventeen. He spent two thirds of the day at the paper's office, even in hellish summer temperatures: sweat and ink transformed his suit into the coat of "a Dalmatian dog" in the words of one of his companions. His train journeys to cover imperial events in Hindu and Muslim territories, with six-month stays that anticipated his famous trips to Japan or South Africa and supplied him with anecdotes and atmosphere for the first stories he published in 1888 in "The Railway Library," the paperback series published by A. H. Wheeler, who thus became his first publishers. Over time, memory would cloak those colonial experiences within the dreamlike, mythical exoticism of novels like *Kim* or *The Jungle Book*.

Both the Hachette and Moreau and Banerjee networks of bookshops followed British models; in 1848, five years before the first French outlet, a similar one already existed in London's Euston Station, the property of WHSmith, probably the first large book chain ever. The company benefited from the railway boom to such an extent that the founder's son was able to use its success as a springboard for an illustrious political career. In parallel, his bookshop was being cloned across the country as big train stations also spread with their concourses large enough to accommodate bootblacks and florists; and there was a progressive refinement of railway travel, which was soon to offer the same luxuries and advantages as ocean liners and hotels. At the end of the eighteenth century and beginning of the nineteenth—as Frédéric Barbier explains in his *Histoire du livre*—London bookshops had already opened up to the street through their shop windows, posters, signs and even announcers or billboard men, who invited passers-by to go in. In fact, the book was now assuming its natural role as a commodity: the list of the remaining titles in the same series or from the same publishers was advertised in the last pages of the book; front pages took on a uniform design to reinforce the identity of a list and innovative illustrations were incorporated; the price began to be printed on the book as a ploy to hook readers or as a publicity device. La Bibliothèque des Chemins de Fer sold books at prices ranging between 0.75 and 2.50 francs. The average price of a book in France fell from 6.65 in 1840 to 3.45 francs in 1870. Series at one franc were created, because the consumption of printed media was multiplying, as was the number of sales points, and those where you could borrow, as were mobile libraries and bookshops that connected up with the nervous heart of the Industrial Revolution like trains. And professional readers: in the nineteenth century there were people who made a living from reading the news out loud or reciting passages from Shakespeare with a flourish. The anachronistic, histrionic Bruce Chatwin did this as a child in Stratford-upon-Avon.

Mobility is the great invention of the nineteenth century. The train changes perceptions of space and time; it not only speeds up human life, but transforms the idea of a network, a network structure, into something that can be explored in its entirety in a few days, even though it is so vast. A whole system reduced to body size. Travellers, who only knew how to read silently, after a period of adaptation, now learn to do so in movement. Not only that: they can also look up from the page, thread together the fragments they have read and imagined with the fragments of life they see through the window (thus preparing for the arrival of the cinema). Lifts appear and allow cities to grow vertically after centuries of horizontal expansion. The heavy furniture of the aristocracy and haute bourgeoisie slims down into light items that allow for house moves. "The street lords it over the room," as Renato Ortiz translates this into spatial terms. The swiftest, most massive migrations in the history of humanity take place. The universal exhibitions of Paris and London, fruits of industrial growth and imperial expansion, are responses to the need to make that supremacy public on the world stage. They are megaphones, monstrous shop windows for the Myth of Progress. Fashion is born, develops at a vertiginous pace that requires mass production, the new consumer society, based on the requirement that everything, absolutely everything must have a sell-by date. Fashion and lightness also shape books: paperbacks, cheap miniatures, discounted titles, bargain boxes, trestles where second-hand books are displayed. All this happens in England and France, in London and Paris, the same contexts where the modern bookshop is created and, alongside, the bookshop chains.

The first Hudson News, with its offering of newspapers and trade books as we know it today, opened in La Guardia Airport in 1987 after a previous attempt in Newark. It now has 600 outlets in the United States. It belonged to the Hudson Group until it was acquired by Dufry in 2008, a Swiss group specializing in duty-free shops. Until his death in 2012, Robert Benjamin Chen was the visible head of the firm that, over decades, had principally focused on the distribution of newspapers and magazines. According to his obituary in the *New York Times,* he received a court sentence in 1981 for bribing the newspaper distributors' union in return for favourable terms. The Hudson Group opened hundreds of bookshops and kiosks in airports, train stations and coach terminals throughout the world and also ran adjacent fast-food restaurants. If the world speeds up in the nineteenth century, the United States is responsible for a second big spurt after the two World Wars. And independent bookshops and book chains—if that polarization is entirely valid and is not challenged in part by an infinite number of intermediate states—develop their structure in the twentieth century, and from the 1950s begin to incorporate the big changes in the consumption of space and time brought by mass culture in North America. The shopping centre that initially imitates a European model (the arcade) is established in city centres, and progressively becomes a suburban phenomenon. And the theme park melds into the fast-food restaurant: in the same year that Disneyland is launched, the first McDonald's franchise is opened, and, with the motel, both connected through the US roadway network, an imperial complex that is duplicated by air routes, the twentieth-century equivalent of the nineteenth-century European rail networks.

Bookshops in the second half of the twentieth century possess the agglutinating character of shopping malls, where the display of books, kindergarten, children's playground, entertainment palaces, restaurants and, gradually, videos, CDs, DVDs, video games and souvenirs cohabit or are neighbours. This vibrant, rather bookish North American model for urban living is copied in large measure by other countries like Japan, India, China and Brazil, and, by extension, everywhere else. The old empires have no choice but to adapt to that hegemonic tendency of a massive leisure offering that ensures the indiscriminate sale of cultural consumer items. In Canada, WHSmith and Coles merged to create Chapters. And Fnac, born in 1954 as a kind of literary club with a socialist ethos, will end up selling televisions and owning some eighty branches in France and more than sixty in the rest of the world. All chains have something in common: what they have on offer is dominated by American cultural products.

THE CIRCULATING LIBRARY.

In his *Atlas of the European Novel 1800–1900*, Franco Moretti has drawn maps of the influence wielded by writers like Scott, Dickens, Dumas, Hugo, Stendhal or Balzac, and of the viral spread on the Old Continent of sub-genres like the sentimental, nautical, religious, Oriental or silver-fork novel (which were sometimes read only in certain regions). This allows him to analyse the logic of the novel form during the nineteenth century as a rendering of two predominant models:

> Different forms, different Europes. Each genre has its geography—its geometry almost: *but they are all figures without a centre.* See here how strange novelistic geography is—and doubly strange. Because, first, the European novel closes European literature to all external influences, it strengthens and perhaps even establishes its *Europeanness.* But then this most European of forms proceeds to deprive most of Europe of its creative autonomy: two cities, London and Paris, rule the entire continent for over a century, publishing half (if not more) of all European novels. It's a ruthless, unprecedented *centralization* of European literature. Centralization: the centre, the well-known fact; but seen for what it really is: not a given but a process. And a very unlikely process: the exception, not the rule, of European literature[. . .] With the novel, then, a *common literary market* arises in Europe. One market: because of centralization. And a very uneven market: also because of centralization. Because in the crucial hundred years between 1750 and 1850 the consequence of centralization is that in most European countries the majority of novels are, quite simply, *foreign books.* Hungarian, Danish, Italian, Greek readers familiarize themselves with the new form through French and English novels, *models to be imitated.*

If we were to apply Moretti's analytical method to the stock of Barnes & Noble, Borders, Chapters, Amazon or Fnac, as he does to nineteenth-century circulating libraries and *cabinets de lecture*, beyond the corresponding percentage of local titles, we would find that the global consumption of fiction is above all the consumption of products from North America or inspired by them. The same strategy that England and France pursued with the novel in the nineteenth century was adopted by the United States, which in Hollywood, and later in television, created a model of audio-visual fiction that was *worthy of being imitated*—just as London and Paris imposed their idea of the bookshop—a way of shaping the space for the experience of family life (with the television at its centre), the experience of watching films (in multi-screen cinemas) and the experience of reading (fusing bookshop, souvenir shop and Starbucks-style cafeteria).

The big North American book chains are consequently the epitome of that way of conceiving the distribution and sale of culture that we constantly mark out with the adjective "big." Because the small chain, the half-dozen bookshops with the same owner and the same brand, may still be capitalized locally, a feature of independent businesses, while the big chains are nearly always transnational conglomerates, where the bookseller has ceased to be simply that, because he has lost that direct—artisanal—relationship with books and customers. The bookseller is a shop assistant or executive director or buyer or personnel manager. Bookshop chains, subject to shareholders and boards of management, trigger series of events that are typical of large enterprises: Waterstones was created in 1982 by Tim Waterstone after he was fired by WHSmith, which, in its turn, bought it in 1999, only to sell it years later to the company that had already purchased its main rival, Dillon's, the branches of which were transformed into Waterstones. Under its new management, Waterstones in Cardiff cancelled a reading by the poet Patrick Jones, after the

Christian Voices Association threatened to boycott the event because the book was "blasphemous" and "obscene."

When I visited London early in 2016 I had the opportunity to interview James Daunt, the managing director of its 300 branches as well as owner of the eight branches of his Daunt chain. When we met in the cafeteria in the Piccadilly Waterstones, I was surprised that the first thing he did was ask me what I wanted to drink, then went to the bar, asked for a coffee and served me with a broad smile. The fifty-two-year-old James Daunt struck me as tall and elegant with a friendly, extremely soothing manner that contrasted with his sharp, incisive eyes. He was appointed in 2011 by the Russian multimillionaire Alexander Mamut, who had just bought the practically bankrupt chain from the HMV Group for 67 million euros. In other words, I interviewed the man who set about saving Waterstones.

"How did things look when you were put in charge of Waterstones in 2011?"

"The chain was bankrupt. Kindle had made big inroads and the market had been reduced by a quarter. What did I do? My first thought was that I must motivate the booksellers, but unfortunately before I could do that I had to sack a third of my staff. My idea was to change Waterstones into a company where I myself would feel comfortable working. It wouldn't be easy, if you think how fixed retail prices are a thing of the past and Amazon can sell books up to 40 per cent cheaper than you can. Consequently the bookseller must make up for the price difference with his humanity, commitment and enjoyment of the synergy existing between the reader, the book and himself. Amazon cannot offer that kind of synergy."

"What were the main changes you introduced, apart from the cutback in staff?"

"Changing a bookshop is a slow process. Hatchards was an historic, very important bookshop, but it was going downhill, and it took three years to reinvigorate it within the Waterstones framework. We are also managing to do that with other shops in the chain: our takings increased from £9 to £13 million last year. The first thing I did was to put my trust wholeheartedly in our booksellers and give them the independence to decide which books they wanted to sell and which they didn't. To that end I was forced to make Waterstones the only chain that didn't allow publishing houses to purchase window or table display space. Previously Waterstones had earned £27 million from selling display space. But it means being pressurized by the publisher and your bookseller isn't free to select books, he isn't the curator of his own bookshop: the job ceases to be stimulating. The purchase of display space creates bookshops that are all the same. The other big change I introduced was in the area of returns. We've gone from 27 to three per cent and my aim is nil."

"The whole system is based on the delivery of new stock and regular returns. You must have a hard time negotiating that with publishers . . ."

"They hated these measures. You have to be courageous if you want to change the publishing world. I met with them and asked them the following: 'Have you got a better idea, because if we don't make changes this business is done for.' They gradually came round. If you are a great publisher, building an important list, you can survive with us; but if you are only interested in novelty, mediocre books you can sell using gimmicks, you will sink."

"What's your relationship with your customers, your readers?"

"The challenge we face is satisfying the most intellectual customer without frightening off the least intellectual. I want taxi drivers to feel at home in all my bookshops. They are people who read a lot, whether it's newspapers or books, and I want them to come into my shops and find what they want to read. I'm not being disingenuous, I know that Waterstones is a middle-class bookshop and my customers in Daunt's have cachet. Each bookshop must get to know its clientele and not attempt to compete with the supermarkets or other establishments that sell books."

"What is a Waterstones bookseller like? A Daunt's one?"

"I hope they will end up being similar. A good bookseller must be friendly, interested in culture and able to communicate that interest, be committed to books and what's more energetic (we mustn't forget it is physical work, too). We want young, well-read staff who choose us because they realize we value a spirit of curiosity and a passion for books. That's why we're changing the design of the space. Whenever I go to Spain I visit La Central, one of my models, like Feltrinelli in Italy."

"You can see that in the warm, welcoming wood on the first floor that recalls La Central in Callao, Madrid. It's clear the same designer is behind these projects: the Argentinian Miguel Sal—"

"Indeed! I would have dinner with Miguel whenever he came to London. He was an intelligent, amusing and provocative man . . . What's more he was an excellent customer, he always ended up buying books like a madman. His recent unexpected death was a great shock."

"What do you think about Amazon's great new idea, opening bookshops that exist physically?"

"I have just returned from Seattle. The bookshop is amazing. The books aren't placed on their side but face up, showing the front cover. There are only 5,000 and they are arranged according to a mathematical pattern; there is no hierarchy of taste, no possible sense that you are discovering anything. They have deconstructed the idea of a bookshop: it would be ridiculous under any other name, but as it is Amazon it's brilliant. Because one should never forget that WHSmith isn't a bookshop and Amazon certainly is."

"Who better to go to war against Amazon than an Amazon?" asked Jan Hoffman in a report on McNally Jackson Books published in the *New York Times*. The warrior would be Sarah McNally, who installed the Espresso Book Machine in an emblematic corner of a bookshop famous for its generosity towards Latin American writers (managed by Javier Molea), for its many activities and its stock of geographically organized books—a machine able to print and bind in a matter of minutes any of the seven million titles in the bookstore-cloud that depends on the tangible Manhattan bookshop.

In a scenario engineered physically by Barnes & Noble and virtually by Amazon, after the closure of hundreds of Borders bookshops, the American Booksellers Association launched the Book Sense and IndieBound campaigns, the keystones of which are a literary prize and a list of the most sold books that only takes into account those purchased in independent bookshops (unlike the *New York Times*, which tots up the sales in kiosks, chains, drugstores and gift shops, as well as the numbers from the publishers themselves, thus often doubling the figures for the same book). In 2010, André Schiffrin commented on that situation in *Words and Money*:

> New York, which in the post-war years had some 333 bookstores, now has barely thirty, including the chains. A parallel development had taken place in Britain, where the Waterstones chain, which had driven many of the independents out of business by its use of huge discounts, was itself bought up by WHSmith. Long known for its commercialism and political caution, WHSmith soon changed its new purchase into a chain focusing on discounted paperbacks.

In his book the publisher uses various labels to differentiate quality bookshops from bookshop chains: "bookstores with a cultural function," "intellectual bookstores," "landmark book-stores," and refers to the protectionist strategies pursued by the French government to guarantee their survival. Years later Hollande's government introduced others. Unlike the Video Club, though never scaling the heights of the Library, the Bookshop wears a halo of prestige, a traditional importance comparable to the Theatre or the Cinema, as a space that must be conserved and developed through state support. This consciousness does not exist in the United States, but it wouldn't be at all surprising if the void left by Borders, rather than being filled by other chains, was occupied locally by new establishments with intellectual ambitions that offered a personal touch and aspired to become centres of culture and future landmarks. Places intensely active on social media, with good web pages, and able to offer printing on demand or nearby printing facilities. Small shops that serve coffee and home-made cakes or offer writing workshops, like those sophisticated wine bars that provide wine-tasting courses. Bookshops that are not dusted by anonymous cleaning agencies but by the booksellers themselves, wanting to remember the precise spot for each of those rare, minority, hand-crafted, out-of-fashion volumes that don't belong in the big book chains and which only booksellers from the family of Beach, Monnier, Yánover, Steloff, Sanseviero, Ferlinghetti, Milla, Montroni or McNally will know how to place on shelves or tables for new books, and thus give them visibility.

XI
Books and Bookshops at the End of the World

What was the first thing I did when I arrived in Sydney? I looked for a bookshop and bought a paperback edition of Chatwin's *The Songlines*, the Spanish translation of which I had read some time ago, and of *Austerlitz* by Sebald, which had just been published in English. The next day I visited Gleebooks and got one of the first stamps on my invisible passport, which at the time (mid-2002) had what you might call a *transcendent* significance for me; I made pilgrimages to cemeteries, cafés, museums, the temples of modern culture I still worshipped. As you will have guessed by this point, I embraced my status as a cultural tourist or meta-traveller some time ago and stopped believing in invisible passports. Nonetheless, I think the metaphor is quite apt and, in the case of bookshop-lovers, could serve to camouflage a fetishist and, above all, consumerist drive, a vice that at times seems too much like Diogenes syndrome. I returned from that trip to Australia with twenty books in my rucksack, some of which I discarded in my various house moves before I read, leafed through, or even opened them.

As I was saying: the next day I went to Gleebooks, though I bought the two key books for that trip in an ordinary bookshop. One must distinguish between the world's great bookshops and emergency bookshops. The latter supply our most urgent reads, the ones that cannot wait, bring light relief on a flight or train journey, allow us to buy a last-minute present, and give us—on the same day it has been released—the book we've been waiting for. Without these emergency bookshops, the others would not exist, would have no meaning. A city must have a range of book outlets: from the kiosk to the main bookshop, there is a whole gamut of modest and middling bookshops, of book chains, best-seller sections in supermarkets, second-hand bookshops, book-shops specializing in film, comics, crime fiction, university textbooks, the media, photography or travel.

I ventured to 49 Glebe Point Road, that colonial-style house with its Uralite porch supported by metal columns, because my guidebook singled it out as the finest Australian bookshop, one that had several times won the prize as the best in the country. It was July 2002 and this book was just one project among many. My notes from that visit, anchored in time, contrast with the book-shop's web page, constantly being updated. "Founded in 1975," I now read in my old handwriting. "Wooden shelves," I read:

An appearance of chaos (there are even books on the floor). The back opens onto a rough-and-ready yard with a few trees. A large amount of Australian, Anglo-Saxon and translated lit-erature. They sell Moleskine notebooks. A mural with the front covers of books signed by their authors. A delightful attic, carpeted like the floor beneath, with lots of natural light, fans and wooden beams, an exposed roof. An edition of Carey's novel on the Kellys, imitating old paper and typography. An up-to-date magazine stand. Literary events are held in the attic. I leaf through a novel about a nineteenth-century prison shot through with absurd humour.

Peter Carey's *True History of the Kelly Gang* was published in Enrique de Hériz's Spanish translation a few months before my trip. In yet another display of his ability to ventriloquize, the Australian writer takes on Ned Kelly in the first person: the orphan, horse-thief, pioneer, reformer, outlaw and policeman. Oedipus and Robin Hood reincarnated at the end of the world. Namely, an interpretation of European myths in a country that, to invent itself as a nation, turned its back on a complex, ancient local culture while simultaneously trying to exterminate or assimilate the indigenous population. Like all Australian bookshops, the layout of Gleebooks demonstrates the unhealed scar that runs across the island continent like a trail of gunpowder: sections called "Aboriginal Studies" and "Australian Studies," because two Australias are thrusting themselves on a single map and each defends its own boundaries.

There are no more bookshops in my archive from that trip to Australia: none of those I visited in Brisbane, Cairns, Darwin or Perth seemed particularly seductive. I found the main titles for my research on Spanish migration to the other end of the planet in museum shops. Ten years later, I visited Melbourne and had the chance to get to know its two main bookshops, which *did* seem memorable: Reader's Feast Bookshop, in whose armchairs I discovered contemporary Aboriginal literature through Tara June Winch, and Hill of Content, without doubt my favourite, as much because of its books as its context. The whole city is articulated through its café cult, so bookshops seem like an appendix to a ritual whose rhythms are totally in step with those of reading. Pellegrini's is an old cafeteria and Italian restaurant, a real Melbourne institution, less than a hundred metres from Hill of Content, and Madame Brussels a sophisticated spot on the third floor of the building opposite. Between vintage antique (the owner speaks dialect to her assistant in the kitchen) and retro modern (only the crockery in Madame Brussels is really old) is where I began to read *Under the Sun*, Chatwin's letters, and *Travels*, the anthology of Bowles' travel writing, both recently published and still not to be seen in my bookshops in Barcelona, yet displayed in the window of one of these bookshops at the end of the world.

Cappuccinos served in Melbourne and the insistence on teatime, the excellent wines and beach huts, pavement cafés and restored arcades can all be seen as a tug-of-war between a Mediterranean, European, and, if you will, international style of life and a degree of resistance to abandoning the British colonial past, the Commonwealth heritage. Just like South Africa: the same cappuccinos, teatime, fine wines, colourful beach huts, the pavement café culture now shared by the majority of countries across the world, the same arcades (and, basically, identical restoration). In the most picturesque in Cape Town, the Long Street Arcade, bookshop and café rub shoulders with antiquarian bazaars and shops selling militaria, a mixture you find in all city arcades in what was once the British Empire.

What was the first thing I did when I landed in Johannesburg in September 2011? Naturally, I asked after the best bookshop. I could not visit it until my last day when, on my way to the airport, I asked the taxi driver to stop and give me sufficient time to get to know it. It was Boekehuis, which specializes in literature in Afrikaans. It is the only bookshop I know that occupies a whole villa, surrounded by gardens and protected by a high wall and a watchtower. A hundred years old, the colonial-style building used to be the residence of the daughter of Bram Fischer, a leading anti-apartheid activist. The fireplaces have been blocked, but it retains a homely atmosphere, the cafeteria is a kind of oasis and the carpets in the children's section welcome storytellers of a weekend. Now that I possess the library that I need and can store books on my tablet, I only buy those titles that can be really useful to me when travelling, the books I cannot easily find in my city and really want to read. So I bought nothing in Boekehuis. Nor in the Book Lounge, the best bookshop in Cape Town.

I had André Brink's *Praying Mantis* in my suitcase. Set in the country's murky dawn, the novel is a rewriting of the true story of the trouble-making Cupido Cockroach, who became a fervent missionary and experienced in his own black flesh the conflicts that were to plague South Africa's future. Both *True History of the Kelly Gang* and several books by J. M. Coetzee employ the same strategy: a manuscript found and rewritten, a dialogue with material from the past. The re-imaginings of the country's troubled beginnings is there in Coetzee's own beginnings as a novelist: the first part of *Dusklands*, "The Vietnam Project," starts with: "My name is Eugene Dawn. I cannot help that. Here goes," and the second part, "The narrative of Jacobus Coetzee," in which J. M. Coetzee figures as translator: "Five years ago Adam Wijnand, a bastard, no shame in that, packed up and trekked to Korana country." *Disgrace* could be translated into Spanish as *Vergüenza, Shame*. Just before my trip to South Africa I had read *Estética de laboratorio*, by Reinaldo Laddaga, one of the few good book-length essays, like *The World Republic of Letters* or *Atlas of the European Novel*, in which the author does not focus on one language or a concrete geographical area but tries to draw a *mappa mundi*: literature cannot be understood if one retains an anachronistic faith in borders. Unlike Laddaga's previous books about Latin American literature, this new title situates the spectre of present-day literature on a wavelength similar to mine (Sebald, César Aira, Sergio Chejfec, Joan Didion, Mario Levrero, Mario Bellatin), alongside other areas of contemporary artistic creation, like music or the visual arts. One chapter analyzes an aspect of *Disgrace* that had escaped me though I had read it several times. In the course of the novel, David, the protagonist, tries to write an opera, the story of Byron in Italy, and the fiction concludes with a desolate image: the character tuning his daughter's old banjo, wondering whether the throat of a dying dog could bring the woeful tone the work requires, sitting on an old chair, under a beach parasol, with jet black, incomprehensible Africa extending as far as his eyes can see, and which does not speak English and is not familiar with the myths and languages of Old Europe. Laddaga argues that this creative endeavour that obsesses David throughout the novel contains the seed of all Coetzee's later books:

pages written from meagre materials, like jottings, diaries, interviews and letters, without the prestige of "the literary," failed essays, attempts to fine-tune a music that refuses to be sublime, where the writer's alter egos appear on the scene and reiterate their inability to elaborate a perfect, rounded story in the twenty-first century.

So similar to Hill of Content or Eterna Cadencia that they could be sisters, the Book Lounge is a charming bookshop with large wooden tables and sofas and a basement with rugs that makes you want to stay on and live there. Its aesthetic is completely classical and therefore familiar, but when I walked around, I confronted an enigma. As I looked at the books, shelf by shelf, I kept finding empty spaces. The first was Paulo Coelho: his novels and self-help books were not there and a small card noted their absence. The second was Gabriel García Márquez. The third, Coetzee. In each case, the same little card with the same message: "Ask for his books at the counter." What could Coelho, García Márquez and Coetzee have in common? The bookseller was chatting to a friend and I was too shy to interrupt, so I killed time taking photographs of the shop and browsing. Finally she was free and I asked her to solve the riddle. And she did: they are the three most-stolen writers. The only ones people steal. So we keep their books here, she said, pointing to big piles behind her. I asked her for Coetzee's. There was not one I did not already have at home, but I leafed through his Nobel Speech again, beautifully cloth-bound by Penguin, which I purchased in the Seminary Co-op years ago. I scoured the edition of *Disgrace* with notes that Penguin Classics had just published and intended for university students for a reference to the aesthetic of precariousness, to the opera that David is writing as the seed of Coetzee's future fiction, or to its poor execution on an out-of-tune banjo in a place only inhabited by dogs. All in vain.

Summertime is the book where Laddaga's intuitions most strike a note. His analysis goes as far as *Diary of a Bad Year*, but it is in Coetzee's latest masterpiece to date where he might have reached a splendid epiphany. A harsh, fictionalized memoir, it is a novel without a centre, without a climax. And yet I remember with peculiar intensity the night John spends with his cousin in a truck, a powerful scene like a whirlwind within a maelstrom despite an apparent indolence, the veneer of non-action. That is the precise moment the reader feels he is at the end of the world. It is a powerful sensation: like crossing Australia or South Africa or the United States or the north of Mexico or Argentina, and suddenly halting after hours journeying through monotonous landscape, stopping at a service station or a village, suddenly being in the middle of nowhere, feeling dizzy at a frontier post where you gaze at the horizon and anticipate the barbarians who never in fact appear, an anguish that prompts the inevitable question: what the devil am I doing here?

In Patagonia I followed the traces of Chatwin like nowhere else on the planet. My copy of the Muchnik edition of his first work thickened out over those few weeks until it became a folder: ridges caused by pencil underlinings were joined by bus tickets, postcards, tourist leaflets, like the ones for the Harberton Ranch or Milodón's Cave. There were two moments when I felt nearest to the author of *Anatomy of Restlessness*: when I interviewed the grandson of Hermann Eberhard ("In the morning I walked with Eberhard in driving rain. He wore a fur-lined greatcoat and glared fiercely at the storm from under a Cossack cap") in Punta Arenas, who told me of the strange visit by writer and biographer Nicholas Shakespeare, who, at one point in the interview, became obsessed with buying his old fridge, since he collected them, and kept returning to the subject of domestic electrical goods until it became the only topic of conversation, and when I walked around Puerto Consuelo to the legendary cave and ended up being chased by a pack of dogs. I had to jump over fences, because the path ran across private property, until finally, by the time I was feeling scared to death, a rough, unkempt guy stepped out of a rusty mobile home that had been converted into a permanent abode and calmed those hellish dogs down. Chatwin, creator of myths: you could not possibly have done everything you describe in your book and yet what an intense feeling of truth radiates from everything you wrote.

What was the first thing I did when I arrived in Ushuaia in the spring of 2003? I visited the Prison Museum and in its souvenir shop bought *Uttermost Part of the Earth* by E. Lucas

Bridges, the story of his life at the end of the world, among the Yahgans (indigenous people), the Onawas (nomad hunters) and his family of British immigrants (owners of the Harberton Ranch, the first on Tierra del Fuego). It is one of the best travel books I have read and the antithesis of Chatwin's story. Bridges pits unity against the latter's fragmentation; and against his superficiality—inevitable given the speed of most memorable journeys—a depth seldom seen in the Restless Tradition. Its author studied the language of the aboriginals, befriended them, established a bridge between Hispanic and Anglo-Saxon cultures that does not even surface in *In Patagonia* as a possibility. Bridges' truth is superior to Chatwin's. Strange but true: literary truth comes in degrees and honesty, unverifiable though it is because with time, facts become ever more elusive, can have a profound effect. A traveller can often see what the native is unable to appreciate, but it is not the same to be a tourist at the end of the world as to have lived there.

I imagine that what I felt on my fleeting stays in Tierra del Fuego, the Cape of Good Hope or Western Australia, that *frisson* of the remote and the finite, must have been similar to what Roman travellers and medieval pilgrims experienced when they contemplated the different ends of the earth with Celtic resonances where Western Europe hurls itself into the sea. After reaching Santiago de Compostela, a university city with an annual bookselling and book-pawning fair which dates back to 1495, the pilgrims would continue for three or four days until they came to Finistère, where on the beach they would burn clothes they had worn over months of wandering before beginning their slow trek home, on foot, as always. If all religions share some things, it is the need for the book, the idea that walking brings one nearer to the gods and the conviction that the world will come to an end. For the ancients that certainty was expressed in physical terms: at a certain point, once a particular frontier had been reached, it was impossible to go any further. We have mapped the most remote corner of the globe and eliminated the mystery of space: all that remains for us to do is to register the end of time.

It has been our fate to witness the demise of the paper book, though it is proving so slow perhaps it will never happen at all. Yet in Bécherel in Britanny, birthplace of the fictional material shaped by Chrétien de Troyes that was so worthy of imitation, a few miles from the French *département* of Finistère, in a single afternoon I visited seventeen bookshops and art galleries linked by ink and calligraphy with the translator François Monti. Bécherel forms part of a spider's web of small bookselling towns that may seem anachronistic, but are very striking. Hay-on-Wye was the first, founded by Richard Booth in 1962; today it has thirty-five bookshops. Small bookselling towns exist in Scotland, Belgium, Luxembourg, Germany, Finland, France and Spain. Before 1989 there was not a single bookshop in Bécherel. Its old textile glory had been reduced to street names: rue de la Chanvrerie (Hemp Street), Rue de la Filanderie (Weaving Street). The imposing merchants' residences speak of the fifteenth, sixteenth and seventeenth centuries, when this area exported the best Britanny linen. The bed and breakfast where we stayed had a distaff and a bookcase full of books. I have never seen so many bookshops with wall-to-wall carpeting.

The houses are old, but the shops selling old books are new and their lack of order is carefully contrived: a retro scenario within vintage architecture. With its two floors and a conservatory adorned with metal sculptures next to the presbytery garden, the Librairie du Donjon is one of the most beautiful I have ever visited. Nonetheless, I find it hard to forget that I am in the midst of a tourist operation. That Bécherel is a book theme park. An old dynamic has been turned upside down: libraries, in total economic crisis, with their collections of games and videos, are more vigorous than ever, yet bookshops are being transformed into museums as part of a strategy for survival.

Or they are disappearing: I have just discovered on the web that Boekehuis shut down in 2012.

Some bookshops, remote if one measures their distance from Barcelona, are to be found at the end of the world. But every single one inhabits a world that perhaps is very, very slowly moving towards its end.

XII
The Show Must Go On

In Venice too I felt that one of the worlds we call a *world* was coming to an end. It was the beginning of December and high tide was daily transforming the Piazza San Marco into a pond with duplicated columns, into a lagoon crossed by wellington-booted tourists, into a shipwreck of metal tables with long legs that liquid reflection changed into metallic herons' legs. It was an opportune moment to pay a visit to Acqua Alta, the place Luigi Frizzo has transformed into one of the world's most photogenic bookshops, with a long gondola stuffed with second-hand volumes in the middle of the central aisle, and a side room that floods several times a year. Planks allowed me to photograph the floor the tide had invaded, part of a city that is adrift, and the stairs Frizzo had built with books gave me access to a beautiful view over the canal. Acqua Alta is not just a bookshop: it is a postcard shop; it is a community of cats; it is a store with boats and baths full of magazines and books; it is a place where you can converse with friendly Venetians who come daily to meet tourists; it is—in the end, above all—a tourist attraction. A notice on the door welcomes you, in English, to the "most beautiful bookshop in the world." When you leave, your memory full of photos, you purchase a bookmark, a calendar, a postcard, at most a history of the city or a collection of travel pieces written by distinguished visitors, and that is how you pay for your entrance to the museum.

Many beautiful traditional bookshops have resisted the tourist circuit or have managed to ignore its siren song. London's

John Sandoe Books, for example, has everything an amateur photographer could desire: its façade unites three eighteenth-century buildings in a single picturesque image with dark wooden windows that reflect the clouds, while inside, on three floors, 30,000 volumes are piled on tables or placed on movable shelves, and stairs up and down connect the poetry or children's basement with other rooms, full of photogenic corners. But the gorgeous body has a soul, I realized when I was about to leave, having leafed through several books without plumping for any. As is my wont, I asked at the till if they had anything on the shop's history. Then Johnny de Falbe—who, I later read, has been working there since 1986 and is also a novelist—began to perform magic. As if it were bait on a hook, he first regaled me with a delightful little book, *The Sandoe Bag*, a miscellany celebrating fifty years. While I was glancing through it, a pamphlet on display behind his back caught my eye: "The Protocols of Used Bookstores" by David Mason, which I bought for £5. We talked about the author, a Canadian bookseller, and suddenly de Falbe disappeared—as any self-respecting magician must at some stage—only to reappear with *The Pope's Bookbinder*, Mason's memoirs, recently imported from Ontario. Before becoming one of North America's great booksellers, he lived in the Beat Hotel, with Burroughs typing furiously in the next-door room, and sought refuge more than once in Whitman's Shakespeare and Company. On his return to Canada he could feel his vocation as a bookseller germinating deep down. I willingly bought this book that I didn't know I wanted for £25. I left Acqua Alta, on the other hand, without buying a thing.

There are two photos of the original Shakespeare and Company in the cafeteria of the Laie bookshop on Barcelona's calle Pau Claris: one of the façade and one of the interior, with Joyce talking to his publishers around a table. To the right you can see dozens of writers' portraits on the wall above a defunct fireplace. It is a miniature gallery, a résumé of the history of literature, an altar to idolatry. Monnier says of La Maison des Amis des Livres: "This bookshop hardly had the look of a shop, and that wasn't on purpose; we were far from suspecting that people would congratulate us so much in the future for what seemed to us precariously makeshift." Sylvia Beach purchased the sofas for her bookshop in the flea market, where, later, Whitman presumably bought his (perhaps they were the same ones!). Steloff transported her few pieces of furniture and the handful of books with which she stocked her bookshop for the first time on a horse-drawn cart. When such an apparently careless aspect lasts for decades, it becomes a stylistic feature and therefore partly a marker. The essence of tourism is that echo from the past, and a classic bookshop, with its veneer of antiquity, must engineer a degree of disorder, an accumulation of strata linked to what cliché identifies with the Great Tradition of Knowledge: an apparent chaos that gradually reveals its orderliness. In the entrance to Acqua Alta you also find locally made products and, as you walk through the different rooms, despite the dust and variegated displays, you start to decipher a system of classification that no bookshop can escape.

The original Bertrand, Lello, the Librería de Ávila, City Lights, the Librairie des Colonnes or Shakespeare and Company have been similarly transformed into museums of themselves fragments of the cultural history they represent, and always have more photographs of writers—as representative icons of the printed word—than of philosophers or historians. That is why people talk, quite unjustly, of *literary bookshops*. With the exception of the ones in Lisbon and Paris, they are also museums of a single bookshop, without branches or clones. The transformation of City Lights into a tourist attraction is practically happening in real time, within the framework of a culture obsessed by distinction and the hectic pace of myth-making

that goes with pop culture. The first Shakespeare and Company was part of the American Express circuit and a tourist-laden coach would stop for a few minutes on rue de l'Odéon so photographs could be taken of the place where Joyce published his famous novel and where Hemingway and the glamorous Fitzgeralds used to hang out. All these shops and others that project a bohemian image and an historical importance appear on lists of the world's most beautiful bookshops that have proliferated over recent years in newspapers and on the web. This has been the case with Another Country in Berlin, a reading club and second-hand bookshop for titles in English. Autorenbuchhandlung, with its refined taste in poetry collections and literary café, and the neighbouring Bücherbogen, five parallel silos dedicated to books on contemporary art and cinema, both in Savignyplatz and under the railway lines: they are the city's best, most beautiful bookshops. The Writers' Bookshop gives material form to a classical ideal of the contemporary bookshop. Book Loans, a *spectacular* ideal: its interior design is fully synchronized with the content of the volumes that comprise its stock. Another Country, on the other hand, simply tries to replicate on a small scale the dusty second-hand bookshop doubling as hostel that gave Whitman such a good return, with a fridge full of beer and American students, the hung-over or night-owls reading slumped on the sofas. Their presence on the lists is the result of two factors: they can be located (and recognized) in English (the journalists compiling these lists also tend to be Anglo-Saxon) and can be summed up by a single image (which is *picturesque* and responds to what we recognize in the paintings, prints and photographs that circulate globally and tend to be repeated; that is, perpetuated through the basic mechanism that regulates tourism and culture: imitation.)

These lists are often headed by a bookshop I have yet to visit, the Boekhandel Selexyz Dominicanen in Maastricht, whose shelves and tables of the latest books are housed in a *spectacular* Gothic structure, a *genuine* Dominican church that was converted in 2007 by architects Merkx and Girod into a shrine to what our era understands as *culture*. They used three metal floors with stairs that ascend, with the columns, towards the top, fully exploiting the height of the nave: upwards to the place of light and the old God. A table in the ironic form of a cross is placed at

231

the end of the nave, in the empty altar space, as if the ritual of communion were solely about reading (consumption moves to the nearby cafeteria). Four years later, the same architects refashioned the original façade with a rust-coloured door that looks like a triptych when open and a box or wardrobe when closed. No doubt it is a masterpiece in architectural and interior design terms, but it's not so clear that it is a fantastic bookshop. It shuts at 6 p.m. and stocks are exclusively in Dutch. Yet this does not matter: style is more important than content in the global circulation of the image. What is picturesque is more vital than the language that leads to reading. The split between the community of readers that allows the bookshop to exist and the tourists who come regularly to photograph it constitutes an essential feature of a bookshop in the twenty-first century. The bookshop became a tourist attraction previously when its historical importance and picturesque condition hit the radar; over recent years, architectural originality, almost always linked to excess, the grandiose and media appeal, has perhaps become a more influential marker than the two traditional ones.

I hope the reader will forgive my abuse of italics at the beginning of the previous paragraph: I wanted to emphasize three concepts: *spectacle*, *authenticity* and *culture*. If, in the twentieth century the building of opera houses, theatres, concert halls, cultural complexes and libraries followed the model of the contemporary cathedral, this same tendency has appeared with force in the domain of bookshops in the present century. The first—now second in most lists, having lost first place when Selexyz was inaugurated—was the Ateneo Grand Splendid that, in 2000, reshaped the interior of a cinema-cum-theatre on Avenida Santa Fe in Buenos Aires dating back to 1919, preserving its dome painted in oils, its balconies, boxes and rails and stage with its dark red curtain. The lighting is dazzling, three circular levels of bulbs create the impression that one is at once inside a monument and in the midst of a spectacle in full swing. An uninterrupted spectacle, where the lead role doesn't fall to customers or booksellers, but to their surroundings. Part of the Yenny chain, the bookshop does not possess particularly remarkable stocks, but guarantees a tourist experience as much for occasional visitors as for locals and keen readers. It offers the experience of a unique place, even though what is on offer is identical to what you find in the chain's others branches. While Fnac can clone itself in the interior of any historical building, converting Nantes Palace of the Stock Exchange into a space identical to the underground area in Barcelona's Arenas Shopping Centre, which on the outside still appears to be—respectively—a neo-classical building and a bull-ring, the Ateneo Grand Splendid displays the uniqueness that is so valued in the symbolic marketplaces of virtual tourism (the image) and of physical tourism (the visit).

I am quite sure that Eterna Cadencia, at one end of Palermo in the same city of Buenos Aires, is a better bookshop, and probably even more beautiful than the Ateneo Grand Splendid. Wooden floors, stately armchairs and tables, excellent stocks set out on shelving that covers the walls, a delightful café on a refurbished patio where all manner of literary events are held, a list published under the same name, and lamps that transport you to Hollywood bookshops. Clásica y Moderna, like the bookshop with that name on Avenida Callao, like Guadalquivir just along the street that specializes in Spanish publishing houses, follows a similar style to the one Eterna Cadencia has recreated in the twenty-first century. We find in all three the same sober style and traditional attention to detail of some of the great bookshops that sprang up in the 1980s and 1990s, like Laie, Robinson Crusoe 389 or Autorenbuchhandlung. And in others that have opened their doors in the last ten years, like the Book Lounge: lots has been written about taste; our era is characterized by the tremendous range.

One can see the project of La Central in Barcelona as a possible migration of the main tendencies from the last quarter of the twentieth century to those of the twenty-first, with the proviso that we not forget the importance of *uniqueness*. The first premises were opened on calle Mallorca in 1996, with a design similar to those I have just mentioned: intimate and human (in step with the reader's body). Conversely, the second, La Central del Raval, established in 2003, synchronizes with Selexyz and Ateneo Grand Splendid in its transformation of an eighteenth-century Chapel of Mercy into a bookish zone, respecting the original architecture and, consequently, the monumental proportions and human-dwarfing high ceilings. However, it has a monastic sobriety, a sense of measure that has disappeared in what might be seen as the third phase of an unpremeditated project: La Central de Callao, in Madrid, established in 2012, completely refurbished an early twentieth-century mansion, preserving its wooden staircase, brick walls, wood and ceramic ceilings, hydraulic-tiled floor and even its painted chapel, and adding, apart from shelves and thousands of books, a restaurant, bar and permanent exhibition of all kinds of objects directly or indirectly linked to reading, such as notebooks, lamps, bags or mugs. Although the ceilings of each of the three floors are relatively low, the extremely high interior patio, with its monumental alphabet soup, brings it in line with one of the main tendencies in our century: a grandeur that allows bookshops to compete with other cultural icons of contemporary architecture.

After it opened, one of the owners, Antonio Ramírez, who embodies the tradition of the nomadic bookseller (the path his life has followed recalls Bolaño's: Colombian by origin, he started in the trade in Mexico City, perfected his training in Paris' La Hune and Barcelona's Laie before starting his own business), published an article (together with Marta Ramoneda and Maribel Guirao) entitled "Imagining the Bookshop of the Future," where he declared:

> Perhaps it is only possible if we locate ourselves in the dimension that cannot be replaced: the cultural density that the material nature of the paper book implies, or rather think of the bookshop as the real space for effective encounters between flesh-and-blood people and material objects endowed with a unique appearance, a unique weight and form, at a precise moment in time.

And he goes on to list the features of that future space that must already be partially present. Ramírez speaks of an architecture for pleasure and the emotion that abolishes all barriers between reader and book, helpfully sketching in a hierarchy of what is on offer, where the bookseller acts as choreographer, meteorologist, hyper-reader or mediator and has to hand the emotional and practical elements that will stimulate the reader's memory and channel his choices—purchases—in the direction that can bring him most pleasure. His emphasis on the bookshop as a summation of concrete physical experiences is of a piece with the architecture and interior design we find in a place like La

Central de Callao, where the spectacular enters into dialogue with our inner selves, the latest fashion complements stocks, the physical feel of paper or card comes into contact with appetites whetted in the bar or restaurant. Unlike other great bookshops of our time, it is plumb in the centre of the city, in a place where crowds walk by, and competes directly with Fnac or El Corte Inglés, aware that—unlike them and their lack of architectural singularity—it can become a tourist attraction if its grandiose tone and picturesque ambience are incorporated into the international circuit of images.

The division of bookshop chains into those that respect the peculiar characteristics of the space that welcomes them and those that impose a single design on all their branches becomes problematic in two Mexican instances: the bookshops that belong to the Fondo de Cultura Económica and the El Pendúlo group. The former is a Latin American chain, with spectacular southern premises like the Centro Cultural Gabriel García Márquez in Bogotá, established in 2008 and spanning twelve-hundred square metres, or the Centro Cultural Belle Époque in Mexico City, which is two years younger and a few metres smaller. While the former and the complex it is part of were created from zero by Rogelio Salmona in the middle of the Colombian capital's historic centre, the Librería Rosario Castellanos is part of the reshaping of the Lido, an emblematic cinema from the 1940s, carried out by Teodoro González de León. It is a dazzlingly white cathedral-like nave where the arrangement of bookshelves and sofas brings to mind a Pharaoh's hieroglyph. The bookshop's ceiling was designed by the Dutch artist Jan Hendrix and represents ancient scripts using vegetation. Naturally, it has a café inside, though it occupies a tiny space.

Conversely, the first El Pendúlo bookshop opened its doors in La Condesa district in the 1990s as a clear fusion of bookshop and café. This would be accompanied by a hybrid concert hall / literary academy in line with cultural centres proliferating at the time throughout the Western world, and anticipating bookshops' primary response to the digital threat: one word symbolizes the mix: *cafebrería*. The bookshop as a rendez-vous point, as a place for business meetings, private classes, events, in a subtly Mexican ambience (tablecloths and plants). Over time, they have opened six premises that maintain a unique style adapted to the features of each new space. In Polanco, for example, the restaurant, bookshop and bar are almost entirely equal in terms of square metres occupied, but the bookcases are instrumental in creating the unifying thread, the overall *tone*, in forging harmony between the different sections with their diverse cultural products: music, cinema, television series, art books . . . In Colonia Roma the wall at the back of the bookshop is called on to provide the interconnecting function, transformed as it is into a hyperbolic bookcase crammed with books bordering the stairs to the first floor and terrace, evoking Patrick Blanc's vertical gardens. In El Pendúlo del Sur there is a huge purple panel that plays on its echo of contemporary art. In Santa Fe we find, in its stead, murals that recall pre-Columbian art and Miró. There is a common corporate image that flirts with cool, memorable individual design features. Clearly the grand bookshop is an important development, interacting with

installations and other features of contemporary design and art that can be seen in those vast places: particularly walls and, above all, ceilings. In addition to those in Buenos Aires, Maastricht, Madrid or Mexico City, the same kind of projects have sprung up this century in the United States, Portugal, Italy, Belgium and China.

The Last Bookstore occupies the former premises of a bank in downtown Los Angeles and has preserved the original giant columns: the counter is made entirely of books and overlooked by a sculpture of a large fish that has also been made from hundreds of books. The old industrial premises in Lisbon's Alcántara district, home to Ler Devagar, have preserved—intact and rusty—the printing press from the old days. At the back is a vast wall packed with books that is continuously flown past by a bicycle with wings that open and shut as if applauding in slow motion. The applause is for a project that is without parallel in the bookselling world. Having had two former shops, one in the Barrio Alto and the other in an old weapons factory, Ler Devagar is now the most widely stocked bookshop in Portugal. It is a limited company with forty shareholders who receive no return on their investment, and enjoy no hope of one, because they have purchased all the books in the main premises and those in other parts of the country. It is a huge library that sells books and encourages one to read slowly. It is also a first-rate cultural centre, where things are always happening. I can't think of a better definition of an ideal bookshop. The white platforms that make up the ceiling in Bookàbar, the bookshop and café in Rome's Palazzo delle Esposizioni, have been set on an incline and punctured as if they were supremacist sculptures. An installation of books hanging on threads from the ceiling dominates the view of Cook & Book in Brussels. In the case of Beijing's Bookworm, a giant orange awning fights the *horror vacui*. Because it is all about humanizing the space, reducing the dizziness provoked by all those square metres between the walls, about camouflaging the height of ceilings that are on a factory rather than human scale.

The majority of these twenty-first-century bookshops have one or two cafeterias, if not a restaurant, harmoniously inscribed in a varied whole where books provide an Ariadne's thread.

Décor, furniture, a children's section disguised as a games room, or the interplay between different colours and textures respond to an emotion-based interior design, the aim of which is to prolong a customer's stay in the bookshop, transforming it into an experience that draws on all the senses and on human relationships. I believe that minimalism is more than a stylistic resource: it can be read as a statement of intent. A hierarchy is established on three levels. At the top is the architecture, almost always propelled by straight lines in a space so huge it ends up imposing itself on everything that exists there but doesn't fill it, to the tiniest letter. At an intermediate level the stairs, picture windows, shop windows, murals, sculptures, period furniture and lights are the protagonists whose function is to try to reduce the intensity of a space that was usually conceived for a different kind of social function and has now been recycled and refurbished. At the bottom comes the display of books, the raison d'être of the whole structure, which can never be as important as they were in the twentieth century when bookshops were made to measure, to fit our hands and eyes, because of the magnificence, the lighting, the art-gallery or vintage-store status of their new abode.

In this way, the bookshop becomes a possible metaphor for the Internet: as on the web, texts occupy a significant but small, limited space in comparison to what is invaded by the visual, and above all by what is so indefinite and empty. As in cyberspace, where things are always happening, and are mostly invisible, a visitor to these multi-spaced bookshops is conscious that stories are being told in the area of children's books, that a poet-singer is performing in the cafeteria, that the new-books table or window display has been changed that morning, that a book launch will begin in a moment, that there is a new dessert menu in the restaurant or literary workshops are about to finish their first session. As in the virtual world, we are witnessing new

forms of socializing, social networks, but the bookish variety clings to personal contact, to the fulfilment of the senses, the only thing the Internet cannot offer us. 10 Corso Como makes its intentions very clear via the "slow shopping" tag at the heart of its spectacular bookshop project. The longer you linger mentally or physically in the shop's atmosphere, the more you buy and consume. Although the Italian chain has premises in Seoul and Tokyo, only the *original* in Milan adds a bookshop to its fusion of hotel, café-restaurant, garden, art gallery, and clothes and design shop. If the centre of gravity is still occupied by a cinema in the Trasnocho Cultural complex in Caracas, opened in 2001 around which are grouped gastronomic, artistic or bookish spaces (the El Buscón bookshop), it is testament to a twentieth-century trend, which in any case never caught on: multi-screen cinemas are usually on the top floor and bookshops are simply just another shop with no particular uniqueness or prestige. In 10 Corso Como the nucleus comprises the restaurant and the hotel, around which we find a couple of cultural outposts that legitimize the cultural activities of the complex as a whole. Its bookshop is generically named: Book and Design Shop, because it has no meaning beyond its glamorous context. In an era when gastronomy is now recognized as an art, culture has broadened its boundaries. This can be seen in tourist experiences that encompass every form of cultural consumption. Something similar has been happening from the inception of modernity: when Goethe travelled through Italy, his visits to bookshops formed part of the *spatial continuum* that shaped every journey alongside churches, ruins, the houses of learned men, restaurants and hotels. Travel and bookshops have always stimulated a love of the marketplace.

Intellectual pleasure fuses with voluptuous delight. Today's bookshops are learning more than ever from the success of shops in contemporary art museums where catalogues are only part of what is on offer, and are usually not even the most significant items alongside jewellery, clothes and industrial design pieces. Strengthened by a minimalist context highlighting the unique facets of every item, objects become a focus of attraction. As in my encounter with that teapot in Beijing, we often find the same T-shirt or cup in another shop, at a lower price, but then it does not enjoy the prestigious aura lent by the Pompidou Centre or the Museum of Modern Art. So it is not *exactly* the same object. If it were only a few metres away, within the framework of the exhibition, we would not be able to touch it, but we can do so in the shop. We can touch and buy everything in a bookshop, which is not the case in a museum or the most important libraries. The profit margins on arty presents are much higher than on books. New book-shops are very clear about how tactile experience adds value to their offerings: premises cannot simply justify their existence as the physical space for electronic sales, they must offer eve-rything that web pages can't.

And that necessarily involves luxury. Because a visit to a bookshop distinguished by its history, architecture, interior design or publishing stock spotlights us as subjects who like luxury, members of a different community to the one that consume culture in shopping centres and the big chains. Paul Otlet, in his *Traité de Documentation* (1934), writes: "Comfort competes with luxury and beauty in sales rooms. A refined atmosphere, comfortable lounges, fresh flowers. Some bookshops like Brentano's, Scribner's or Macmillan are real palaces." Megalomaniac bookshops have existed at the very least from the eighteenth-century Temple of the Muses. Policy in eighteenth-century salons was regulated precisely by refinement, by an aristocratically refined taste. With the advent of democracy, the dream of the troubadours is multi-plied exponentially: whether readers belong to the most excellent communities of their era depends on their culture, education, artistic insights and not on their acquisitive power or blood. Nevertheless, it is true that if one wants to be able to evaluate and interpret the architecture, design or stock of spectacular bookshops it is a costly educational process. Not everyone can afford the trips that allow one to explore the high-profile places touted in tourist guides. As in all tourist scenarios, different levels of awareness, intellectual insight and class fictions coexist: as many as the brains and gazes browsing there at a particular moment.

In *A Life of Books*, Joyce Thorpe Nicholson and Daniel Wrixon Thorpe make it very clear that Australian booksellers in the 1970s were aware of how crucial it was for their premises to project what the authors called "a trendy appearance." They mention Angus & Robertson in Sydney, whose owners decided to paint each floor a different colour when they moved to new premises; the Angus & Robertson, in Western Australia, that up and went to a period hotel and tavern and started a campaign based on the coupling of "books and beer"; and the Abbey's

Henry Lawson's Bookshop, in the basement of the Hilton Hotel in Sydney, with its black wood bookcases and impressive offer of "any book published in Australia." There are many other precedents for the spectacular bookshop waiting to be disinterred in libraries, newspaper collections and personal reminiscences. Two nineteenth-century stations that were converted into bookshops still exist: Barter Books opened its doors in 1991 in Alnwick, Northumberland, and four years later Walked a Crooked Mile Books did so in Philadelphia.

The revamping of hotels, railway stations, cinemas, palaces, banks, printer's, art galleries or museums as bookshops is a constant over recent decades and has accelerated in the twenty-first century. In a new historic context in which recycling has taken on a new meaning, in which culture has been digitized and, above all, in which the existence of all that is real is—simultaneously—physical and virtual, these cathedrals to the written word acquire a deeply capitalist, religious-cum-apocalyptic significance that also reveals unprecedented artistic ambition. The impact of the spectacular is decisive on both fronts. Via El Pendúlo's web page you can make virtual visits to each of its *cafebrerías*. Google Images and other platforms are awash with photographs of the world's most beautiful, most interesting, most spectacular bookshops. For the first time in the history of culture these bookshops have immediate access to the international tourist circuit, markers gather pace and generate immediate contagion—at a stick-and-paste rate—on web pages, social media, blogs and microblogs, all of which create a desire to visit, to get to know, travel and photograph without any recourse to History or the participation of famous writers or acclaimed books. The image of a church, railway station or theatre transformed into a bookshop is more compelling in the new logic of tourism than the hundred thousand books in the picture or their ten billion words.

XIII
Everyday Bookshops

J. R. R. Tolkien's first poem, "Goblin Feet," was published in a poetry collection by Oxford's Blackwell's Bookshop, which cancelled his debt in exchange for the advance on his rights, and because he had been a regular customer in that establishment, founded in 1879 by Benjamin Henry Blackwell, which was later transformed into a publishing concern by his son Basil, the first member of the family to go to university and the publisher of the author of *The Lord of the Rings*. As the business grew and became a chain, each branch attracted its parishioners, its usual suspects, its congregation, people who chose Blackwell's in Edinburgh, Liverpool or Belfast as their everyday bookshop.

If you hunt around the main premises in Oxford, it is still possible to imagine how the few square metres where the business was created gradually increased until several houses were transformed into a single monster of a house. As you enter, to the left, a nineteenth-century fireplace and wooden beams are the archaeological traces of the original establishment. If you ask, you can visit the reconstructed founders' office next to the fireplace on the floor above, where pipes, spectacles and letter-openers are laid out on the table as if they had been left there only a few hours, and not a whole century before. The successive owners of Blackwell's bought up all the flats in the building as business expanded. The most recent, definitive extension is a huge basement at the back that occupies the space under Trinity College Gardens. It has its own name: the Norrington Room. It is an Olympic swimming pool of shelves and books. In the 1960s and 1970s, during the frequent power cuts, it relied on kerosene lamps that ensured people could still read, whatever the obstacle. I imagine those readers as if they were marooned in a post-nuclear bunker. From above, despite the rectangular shape, it looks like a giant brain. This is what it is: the brain of a collective intelligence, which is in fact what its eighty employees are for the most part; what Oxford University is, expanding exponentially and intellectually, just like its best bookshop.

The last time I was in Berlin, before I went to photograph the decomposing remains of the Karl Marx Bookshop, I bumped into César Aira, quite by chance. We went to the nearest cafeteria and chatted for a while about the latest literary titles out in Argentina. "We would meet almost daily," he said halfway into the conversation, "in the International Argentina, Francisco Garamona's bookshop, Raúl Escari, Fernando Laguna, Ezequiel Alemián, Pablo Katchadjian, Sergio Bizzio and other friends." Dominated by a sofa and a small table for your glass of wine, the site of the Mansalva publishing house is probably the only bookshop in the world where you can buy most of Aira's books, even in translation, although naturally there will always be ten or twenty that not even Garamona can supply. One of those places where the ways of a past era have become established. Like the Ballena Blanca, Alejandro Padrón's place in Mérida, Venezuela, where university teachers like Diómedes Cordero and writers like Ednodio Quintero meet daily to talk about the country's great poets, about Japanese literature or

political issues in Spain or Argentina while they prepare the next edition of the famous *Bienal de Literatura Mariano Picón Salas,* which was to inspire the exploits of Aira and an army of Carlos Fuentes clones in *The Literary Conference.* Because literature is polemical, and is about the future and about books to imagine.

"In the afternoon our bookshop seemed more like a club where scientists, literati and artists met up, talked, to find relief from the prosaic nature of daily life," wrote Mikhail Osorgin on the subject of the legendary cooperative in Moscow, The Writers' Bookshop. Although conversations about literature are as old as Western culture, it is, of course, in the seventeenth and eighteenth centuries that they would become institutionalized as literary *conversazione.* It should be no surprise then that this coincided with the moment when bookshop and café began to fuse into a single organism, as Adrian Johns has observed in *The Nature of the Book.* Apprentices were part of the family and the boundaries between private space and public business were not at all obvious, so that the presence of armchairs, seats and sofas where one could enjoy reading while drinking was often due to the fact that they belonged to the bookshop owner's house. Since then, many booksellers have developed salons and literary conversations that double as debates about culture and buying and selling sessions: "The foremost example of 'amphibious mortal' was surely Jacob Tonson. Among aristocrats he looked like a bookseller; among booksellers he appeared an aristocrat." The confusion between private and public life parallels the confusion between bookshop and library. In his diaries, Samuel Pepys writes of bookshops where "seats were available so customers could read for as long as they wished." And booksellers themselves in the eighteenth century were the driving force behind lending libraries that were much more democratic than literary societies and the only way in which artisans' apprentices, students or women could have access to literature without incurring the huge expense of a book. One could even say that, despite appearances, bookshops have never really been sure about their real boundaries.

I have found a haven in many of my travels, fleeting homes far from the home I did not actually possess, and found refuge in their ambiguous nature. I remember my daily visits to Leonardo da Vinci's basement during my stay in Rio de Janeiro, to the Seminary Co-op when I lived in Chicago and to the Book Bazaar in Istanbul for as long as my foolish haggling lasted trying to secure that Turkish travellers' book, to the Ross bookshop in Rosario on every one of my sojourns in that city with a river without shores, even though it was in the nearby premises of El Ateneo that I found the complete works of Edgardo Cozarinsky, and, in its café, where I read *Rinconete y Cortadillo* and *El licenciado Vidriera*. Since re-establishing myself in Barcelona, whenever I escape to Madrid, as well as visiting La Central in the Reina Sofía art museum, I try to have a coffee in Tipos Infames, a bar and gallery in step with the latest trends in international bookshops; I go to say hello to Lola Larumbe, who manages Rafael Alberti in such a charming, professional manner, a bookshop designed by the poet and painter in 1975, where water seems to swirl over the basement; I visit Antonio Machado, in the basement of the Circle of Fine Arts, with its delightful selection of books from small Spanish bookshops and by whose cash register over the years I have discovered the main studies of bookshops that I've used for this essay. I go to Naples twice a year and am duty-bound to visit Feltrinelli's in the Central Station and the Librería Colonnese, on via San Pietro a Majella, surrounded by churches, artisans making Nativity cribs, remains of ancient walls and altars dedicated to San Diego Maradona.

There is no doubt that a bookshop is much more hospitable when, as a result of repeated visits or coincidence, you strike up a friendship with one of the booksellers. When I lived in Buenos Aires and Rosario and had to leave the country every three months, I used it as an opportunity to explore parts of Uruguay by sea, land

and river. Every one of my journeys ended in La Lupa, the bookshop where one of its owners, Gustavo Guarino, gave me leads into Uruguayan literature on every visit: only by travelling to the place where things happen do you find access to what resists visibility on the Internet. One of the pleasures that awaits me in Palma de Mallorca is La Biblioteca de Babel, where I can lose myself in its essay and fiction sections, and Literanta, where critic and cultural activist Marina P. De Cabo stands behind the counter; it was she who discovered me when I became interested in the work of Cristóbal Serra. For years I visited La Central de Raval in Barcelona on Fridays, knowing that César Solís would be there to recommend the latest titles from Latin America, or supply me with the latest book by Sebald or on Sebald to be published in one of the main European languages. Ever since he moved to Madrid, I now go to Damià Gallardo, in the Centre for Contemporary Culture's Laie bookshop, who solves my problems as a reader. Because every good bookseller must be something of a doctor, chemist or psychologist. Or barman. Francisco, Alejandro, Gustavo, Marina,

César and Damià form part of my own bookseller tradition, the restless tradition of habits you take up again as soon as you arrive in distant cities where you once lived.

Austerlitz, the hero of W. G. Sebald's novel of that name, experiences the most decisive moment of his life in a second-hand bookshop near the British Museum, one that is owned by a beautiful woman whose name is a pure haven: Penelope Peacefull. While she is solving a crossword puzzle and he absent-mindedly flicking through architectural prints, two women are talking on the radio about "the summer of 1939, when they were children and had been sent to England on special trans-port." Austerlitz's mind and body are invaded by a kind of trance: ". . . and I stood there as still as if on no account must I let a single syllable emerging from that rather scratchy radio escape me." Those words allow him at a stroke to recover his own childhood, his own journey and arrival in England escap-ing from a Europe in flames, his own exile: years his memory had erased completely. In a bookshop, he suddenly remembers who he is, from which Ithaca he has come.

Childhood and especially adolescence are periods when you fall in love with bookshops. I spent so many Saturday afternoons browsing among the shelves of Rogés Llibres, the ground floor of the Garden City of Mataró that had been converted into a sec-ond-hand bookshop, that I am quite unable to shape them chronologically or date them. I am sure those sessions only took place on the weekend and on holidays, because during the school term I went in the opposite direction, towards the city centre. On my way I would go into the Caixa Laietana Library, where I read all the Astérix, Obélix and Tintin comics and took out all the *Alfred Hitchcock and the Three Investigators* and Sherlock Holmes novels, and on my way home at dinnertime would pop into Robafaves, which I discovered much later was a cooperative and one of the most important bookshops in Catalonia, where a book was launched nearly every evening, and I'd listen as if in church or class to those words that, although they were there between mouth and microphone, objects as tangible as the books surrounding me, sounded remote, an incomprehensible babble, completely disconnected from my firm desire to be a writer.

When I was fourteen or fifteen I accompanied my father on his visits to homes in a neighbourhood in Mataró, next to the Central Park, the velodrome and municipal swimming pool, where as a child I saw peacocks, races and cyclists and my own body diving into the water as if I were not scared of all those litres of blue chlorine. After an eight-hour day at the telephone company, my father worked as an agent for the Readers' Circle. First he would distribute the new catalogues and then we picked up the cards with the orders from the subscribers and processed them and a few weeks later,

when all the books had arrived at our house, my mother helped us organize them by street and finally we took them to their new owners and collected payment. Some customers forced us to go back two, three or even four times, because they never had the 950 or 2,115 pesetas their order cost. Conversely, others bought five, seven or nine books every two months and always had the 10,300 or 12,500 pesetas ready because they were expecting us and so much wanted to start reading. I suppose it was in one of those family flats or flats belonging to old ladies or single men, complete strangers, that I first saw well-stocked private libraries and decided that, one day, when I was a writer, I would have one too. The first aspiration was too abstract to be more than pie-in-the-sky; the second, on the other hand, assumed a tangible form that, like girls' bodies, was pure desire.

In *Auto da Fé*, Elias Canetti writes that a child is at the mercy of the wares of any old tradesman dealing in books as soon as they can read and walk. He concludes that young children should be brought up in private libraries. He is probably right, because I cannot recall a single book I bought in Rogés Llibres or in Robafaves that changed my life: the great reads in my life came later (or simply: late), when I had moved away from Mataró. Nonetheless, Robafaves is the most important bookshop in my life because I was introduced to something I had glimpsed in those private houses there: a life in which books played an intimate part. In *2666*, Amalfitano muses that a book probably came to him via Laie or La Central. I could say the same about most of my library, a third, perhaps, to which one should add the "restless" titles bought in Altaïr and the comics purchased in Arkham. The other two thirds come from my travels and what publishers' publicity departments send me. I have sent dozens of boxes from Rosario, Buenos Aires and Chicago: libraries and the nomadic spirit are always wedded in my mind. My own experience of cities is shaped by the intersection between strolling and bookshops, so most of my usual itineraries cling to certain shops as my personal hubs or stops. Streets, book-shops, squares and cafés represent the routes of modernity as well as settings for two essential acts: conversation and reading. While lit-erary writing—which, until a few decades ago, was still taking place at café tables, has been retreating into private space, or at best to libraries, talks and readings, premeditated or chance encounters, and the diary, novel or magazine—continued to be articulated in the social sphere of metropolitan life. Blogs and social media allow you to exchange data and ideas in the cosmopolis, but your body continues to tread a local, domestic topography.

As we read in "The Journey of Álvaro Rousselot," one of the stories in *The Insufferable Gaucho,* Bolaño thought that the bookshops of Buenos Aires and their contents had lives of their own. In other words, it is not only the movement of readers' bodies that threads together the different bookshops in a city, books themselves shift and wander, open lines of escape and create itineraries. That was the idea that inspired the Barcelona theatre director Marc Caellas when he decided to adapt Robert Walser's *The Walk* into a walk around the Argentinian capital. Its pages were suddenly incarnated by an actor, a stroller, who wanders round the various emblematic spaces in the modern city as happens in the story. Of course, one is the bookshop:

As now an extremely splendid, abundant bookshop came pleasantly under my eye, and I felt the impulse and desire to bestow upon it a short and fleeting visit, I didn't hesitate to step in, with an obvious good grace, while I permitted myself, of course, to consider that in me appeared far rather an inspector, or bookkeeper, a collector of information, and a sensitive connoisseur, than a favourite and welcome, wealthy book buyer and good client. In courteous, thoughly circumspect tones, and choosing understandably only the finest turns of speech, I enquired after the latest and best in the field of belles-lettres [. . .] "Certainly," said the bookseller. He vanished out of eyeshot like an arrow, to return the next instant to his anxious and interested client, bearing indeed the most bought and read book of enduring value in his hand. This delicious fruit of the spirit he carried carefully and solemnly, as if carrying a relic charged with sanctifying magic. His face was enraptured; his manner radiated the deepest awe; and with that smile on his lips that only believers and those who are inspired to the deepest core can smile, he laid before me in the most winning way that which he had brought.

I considered the book, and asked, "Could you swear that this is the most widely distributed book of the year?"

"Without a doubt!"

"Could you insist that this is the book one has to have read?"

"Unconditionally."

"Is this book also definitely good?"

"What an utterly superfluous and inadmissible question."

"Thank you very much," I said cold-bloodedly, left the book that had been most absolutely widely distributed, because it had unconditionally to have been read, where it was, and softly withdrew, without wasting another word.

"Uncultivated and ignorant man," the bookseller shouted after me, for he was most justifiably and deeply vexed.

This is the walker created by the Swiss Walser, in some bookshop in Boedo speaking with an Argentinian accent, making fun of conventions, of literature tied to sales figures, of the absurdity of the world of culture, being directed by a Catalan. Marginal centres and central margins, abolished frontiers, translations, changes of city, quantum leaps, transcultural interactions: welcome to any bookshop.

The same relationship between periphery and centre that I experienced quite unawares when I visited Rogés Llibres and Robafaves, as if they were mazes, second-hand and antique book-shops and shops for the latest titles, can also be established between bookshops in the centre of Barcelona and those on the city's fringes. Gigamesh was the first shop I entered in Barcelona and I soon started to explore others selling comics, science fiction and heroic fantasy that surrounded it and still do, like a plague of aliens pro-liferating over time, in the vicinity of the Paseo de San Juan. The area of that impossible centre occupied by Laie, Documenta, Altaïr, Alibri and La Central, and so many others, is small and walkable. Until the end of 2015 you only had to cross El Born, a district without bookshops, to reach La Negra y Criminal that Paco Camarasa ran for almost fifteen years in a backstreet of Barceloneta. Now both districts are orphaned. Bookshops imitate the neighbourhoods that welcome them: this place could only exist among those fishermen's houses, and in Gràcia a mere fifteen min-utes' walk from the Arco de Triunfo, Taifa and Pequod are unim-aginable without the context of a locality, a context of nearness. Camarasa and José Batlló, the Alma Mater of Taifa (now run by his heirs, Jordi Duarte and Roberto García) are two of the key indi-viduals in Barcelona's world of books that finds its originating myth in the pages Cervantes devoted to it in *Don Quixote* and which has always negotiated the city's literary bilingualism. Taifa has been the pre-eminent bookshop north of the Diagonal since 1993, as La Negra y Criminal is south of the Ronda Litoral. Batlló is a poet, publisher and legend. He is famous for his culture, for being a great friend to his friends and for his skirmishes with cus-tomers whom he is capable of scolding, depending on the titles they buy. Those he has sold most over the last twenty years are *Hopscotch* and *City of Marvels*. Second-hand books are in cell-like spaces at the back, as if to remind us that it is normal for novels and essays to cease to circulate, for publishers to shut down, for us to be forgotten. Similarly Pequod, the belly of the whale that, as they say, was born yesterday, sells both new and second-hand books, because we live in hybrid times. Despite its small surface space, this bookshop transforms itself into a gallery for micro-exhibitions, an area for conversations about Italian literature and an aperitif bar at the weekends, and spreads itself around social media, because nothing new exists solely in the world of what we can touch.

In a second circle—orbit within an orbit—other Barcelona bookshops have vied for recognition over the last few years. I am thinking, for example, of +Bernat, the bookshop and restaurant on calle Buenos Aires, next to the Plaza Francesc Macià, which is managed by Montse Serrano and defines itself as a "cultural store," and is the favourite haunt of Enrique Vila-Matas since he switched neighbourhoods. Or of Llibreria Calders in calle Parlament in the Sant Antoni district, with its piano and agenda forever in flames. Or of Nollegiu, in Poblenou, that Xavi Vidal has turned into an important cultural centre. Fortunately, they are not the only bookshops that have been generating urban interest far from the city centre. Because, although hubs where bookshops are concentrated have heritage value, like the popular Port'Alba that Massima Gatta has called "the Charing Cross Road of Naples," or Amsterdam's elegant Het Spui and adjacent streets, a democratic city is a network of public and private libraries and small and large bookshops: a dialogue between readers who live in multiple centres and various peripheries.

My strolls often lead me to calle Llibreteria, the ancient Decumanus in Roman Barcino, where you now find the Papirvm artisan shop and La Central in the Museum of the History of the City, one of those places—like the bookshop in the basement of the College of Architects—where Barcelona archives its own memory. The brotherhood of Sant Jeroni dels Llibreters was founded in 1553. If St Lorenzo, one of the

Church's first treasurers, is held to be the patron of librarians because of his work classifying documents, the austere St Jerome, one of the Church's first ghost-writers (he wrote Pope Damaso I's letters) is held to be the patron of translators and booksellers. St Lorenzo, the man who some legends identify as the mysterious individual who hid the Holy Grail to protect it against the wave of violence that also ended his own life, died a martyr grilled to death on the outskirts of Rome: on August 10 every year the reliquary that contains his head is exhibited in the Vatican, I am not sure whether it is venerated solely by librarians. St Jerome, on the other hand, after a period as an outstanding

translator, went to Bethlehem in self-exile, lived in a cave and devoted his time to attacking in his writings the vices of Europe textually and beating himself with a stone in acts of penitence. He usually appears in the iconography with the Vulgate, the Bible that he translated into Latin from Hebrew—though he was an expert in ancient Greek, too—open on his desk, a skull as a symbol of *vanitas* and that stone which rumour-mongers say he used as a kind of translation dictionary that had yet to be written: he beat himself and God revealed to him *ipso facto* the Latin equivalents of the Hebrew original.

Your city enters its bookshops through their windows and customers' footsteps, a hybrid space that is neither wholly private nor wholly public. The city walks in and out of its bookshops, because one cannot be understood without the other, so the pavements outside Pequod or La Negra y Criminal are always crammed with people on a Saturday afternoon or Sunday morning, drinking wine and eating steamed mussels, and books on Barcelona find their way into every city bookshop, which is the space where they naturally belong. And when they start to get dog-eared, the novels, essays, biographies and books of poetry that citizens have man-handled and owned return to the city's stalls, to the Mercado de San Antonio, to the second-hand bookshops or that arcade with books and a Uralite roof that was at the back of Los Encantes where passers-by turned out to be collectors, antiquarians and rag-and-bone men.

If the metropolis vamps up its bookish dimension on Sundays in the Mercado de San Antonio or on the days Los Encantes is open, there is one day in the year when the city reproduces in its every corner that sensation Don Quixote took away with him: the city breathing the printed word. The Spanish Day of the Book was the brainwave of a Valencian, Vicente Clavel, who had established himself in Barcelona as the youthful owner of the Cervantes publishing house. From the Chamber of Books and with the support of the Catalan Labour Minister, Eduard Aunós, he gained recognition for his project through a royal decree in 1926 in the middle of the Primo de Rivera dictatorship. Although the idea was to encourage Spanish book culture at all levels of the administration, so that every library and city would participate in one way or another in the festivities, from the very first it became polarized between mass celebrations in Barcelona and institutional, academic events in Madrid. Guillermo Díaz Plaja wrote these words in an article after Clavel's death:

> Almost half a century later the decree remains in place, with a single important change—the decree of September 7, 1930—which switched the date of October 7 that was originally agreed—two days before the certified baptism of Cervantes—to

April 23, the day of his death. This historically exact date meant that the Day of the Book in Barcelona coincided with the day of St George, the patron saint of Catalonia. When Don Gustavo Gili pointed that out, Clavel immediately retorted, "It doesn't matter. The roses for St George will flower for ever. The only risk we run is that Cervantes will be forgotten." The years gone by have shown that it was to be a happy marriage of both commemorations in the legendary Barcelona festival. The city of the Counts is without a doubt the vanguard of the Peninsula in terms of the breadth and popularity of the Day of the Book.

The year 1930 was when publishers began to launch new books in Catalan on the Dia de Sant Jordi and the general public responded with enthusiasm, while Madrid took the first steps to organize its Book Fair on another date and the rest of the country also gradually forgot Cervantes' Day. The Civil War paralyzed the publishing industry and Francoism banned Catalan and eliminated the Book Chambers by replacing them with the National Spanish Book Institute. The Day of the Book did not start to become important again in Catalonia until the 1950s. In 1963 the opening address was given by Manuel Fraga Iribarne, the Minister for Information and Tourism, who defended the need to promote literature in the Catalan language. The front page of *La Vanguardia Española* on April 23, 1977 (15 pesetas), together with a photograph of a street packed with people, reproduced in Catalan the following lines by Josep Maria de Sagarra:

> The rose has given him joy and pain
> and who can say how dearly he loves it;
> bringing more blood to his veins
> to defeat all the dragons in the world.

Thanks to an initiative taken by the First Latin American Congress of Book Associations and Chambers, from 1964, April 23 became the Day of the Book in every country where Spanish and Portuguese are spoken and from 1966 it has been International Book and Authors' Rights Day. Perhaps because not only Cervantes and Shakespeare died on that day but also other internationally known writers, like Inca Garcilaso de la Vega, Eugenio Noel, Jules Barbey d'Aurevilly and Teresa de la Parra.

I love to visit my favourite bookshops on the days prior to Sant Jordi: I buy all my books then and during *la diada* I simply like to stroll and observe, "like a better sort of tramp, a vagabond and pickpocket, or idler and vagrant," as Walser says. Like all self-respecting writers and publishers, I use these walks to check whether my books are there or not and to put them in the proper place on the shelves of my everyday bookshops. And in those where they are absent. Even in the book section in El Corte Inglés. Even on the second floor in Fnac, in the middle of the city, where I imagine many of the young sales assistants, with their BAs, MAs or PhDs in literature would have been great booksellers in another—no doubt better—world, or perhaps already are in this one, which although in crisis is the only one we have.

EPILOGUE
Virtual Bookshops

Over the first few months of 2013, I watched how a book-shop that was almost a hundred years old became a McDonald's. Of course, it is an obvious metaphor, but that doesn't make it any less shocking. I am quite sure that the Cat-alònia, the bookshop that opened its doors on the edges of Plaza de Cataluña in 1924, was not the first to be transformed into a fast-food restaurant, but it is the only time I have personally witnessed such a metamorphosis. For three years I walked past the glass door in the morning and sometimes went in to take a look, buy a book, make an enquiry until suddenly the shutters stayed shut and someone stuck up a precarious notice, barely a page long, which read:

> After over eighty-eight years of being open and eighty-two years of activity at Ronda San Pere 3. After surviving a civil war, a devastating fire, a property dispute, the Llibreria Cat-alònia will close its doors for good.
>
> The severe crisis in the book trade has generated a slump in the sales of books over the last four years that has made it impos-sible for us to continue in these circumstances and conditions.
>
> It has been very sad, difficult and painful to take this irrevocable decision. We have tried to find every possible solution, perhaps too late in the day, but either they didn't exist or we couldn't find them.
>
> Nor could we have prolonged this situation, because we wanted to ensure that the business closed in an orderly way

and met all its obligations. If we had continued any longer, the end would have been much worse.

As we make this decision public we would also like to remember all those who have worked throughout the years in the Llibreria Catalònia and the enterprises that depended on it, especially the Selecta publishing house, and also all our customers—some over decades and generations—and our authors, publishers and distributors. Jointly they have allowed the Llibreria Catalònia to make an important contribution to the culture of Catalonia and Barcelona.

Now and in the future, in all the forms that the dissemination of culture will take, there are and will be individuals, associations, collectives and enterprises ensuring the survival of literature and written culture in general. Unfortunately, the Llibreria Catalònia will not be part of that future.

Miquel Colomer, Director, Barcelona, January 6, 2013

Day after day I was witness to the disappearance of books, to empty shelves, to dust, that great enemy of books, books that were no longer there, only ghosts, memories, books that were gradually being forgotten until one Wednesday there were no shelves for them to be on, because the premises were emptied, filled with workers who yanked out the bookcases and the brackets and the place was all noise and drilling, a din that shocked me for weeks, because I had been used to the silence and cleanliness it had emanated for years; when I walked past that same door, I was met with clouds of dust, carts loaded with rubble, with debris, the gradual transformation of the promise of reading, the business of reading into the digestion of proteins and sugar, the fast-food business.

I have nothing against fast food. I like McDonald's. Indeed, I am interested in McDonald's: I search one out on most of my trips, in order to try the local specialities, because there is always

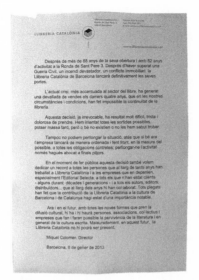

a breakfast or a fajita or a hamburger or a sweet that is the
McDonald's version of one of the favourite dishes of the locals.
However, that didn't make this supplantation any less painful.
For months, every morning I watched the destruction of a small
world, occupying that same space like an ambassador from
another world, and in the afternoons I read about reading and
finished writing this book.

There is a traditional, multicoloured bookshop in Turin
called La Bussola. All bookshops are compasses: when you study
them they offer you interpretations of the contemporary world
that are more finely tuned than those provided by other icons or
spaces. If I had to choose another bookshop to explain partially—complete explanations do not exist—the schism within
the book trade in our era it would be Pandora's in Istanbul. It
has two well-stocked premises, one opposite the other: one
exclusively sells books in Turkish; the other, titles in English.
The prices in one are in Turkish lira; those in the other in dollars. Pandora makes a symbolic reality explicit: all bookshops

live between two worlds, the local and the one imposed by the United States, traditional business (of a local sort) and the one in huge shopping centres (chains), the physical and the virtual. This metaphor is not as obvious as the one afforded by an old bookshop, a classic, vintage bookshop, a bookshop that was founded by Josep López, Manuel Borràs and Josep Maria Cruzet, survived the wintry bunker of a dictatorship and systematic harassment by a real estate company, after fierce political and moral resistance yielded to the cold, implacable, abstract rules of economics, shut down its premises, a few metres away from the Apple Store, two hundred metres from Fnac, opposite El Corte Inglés, and was transformed into a McDonald's. In effect, the Pandora metaphor is more oblique but more hopeful, because it leads to survival rather than closure. All bookshops are divided into at least two worlds and are forced to consider other possible worlds, and I write that without a scrap of naivety.

Green Apple Books—as Dave Eggers recalls in his chapter in the anthology *My Bookstore*—is lodged in a building that has survived two earthquakes that brought turmoil to San Francisco in 1906 and 1989; perhaps that is why one experiences between its shelves the feeling that "if a bookshop is as unorthodox and strange as books are, as writers are, as language is, it will all seem right and good and you will buy things there." I bought there a short bilingual book, published by a Hong Kong poetry festival, the English title of which is *Bookstore in a Dream*. Four lines about the bookshop as a quantum fiction really caught my attention: its multiplication through space, its mental realm, its existence in parallel universes on the Internet, a compulsive survivor of all earthquakes. If Danilo Kiš' narrator dreams of an impossible library that contains the infinite *Encyclopedia of the Dead*, Lo Chih Cheng dreams of a bookshop that cannot be mapped out. A bookshop, like any other, that is soothingly physical and horribly virtual. Virtual because digital, or mental, or because it has ceased to exist. A bookshop that is born, like Lolita in Santiago de Chile, like Bartleby and Company in Berlin or Valencia, like Librería de la Plata in marginal Sabadell. Like Dòria Llibres, which has filled the space left by Robafaves in another small Catalan city, my Mataró: at what point do projects become completely real? Bookshops of the memory, gradually invaded by fiction.

Like the shop run by the wise Catalan in *One Hundred Years of Solitude*, who came to Macondo during the Banana Company boom, opened his business and began to treat the classics and his customers as if they were members of his own family. Aureliano Buendía's arrival in that den of knowledge is described by Gabriel García Márquez in terms of an epiphany:

> He went to the bookstore of the wise Catalonian and found four ranting boys in a heated argument about the methods used to kill cockroaches in the Middle Ages. The old bookseller, knowing about Aureliano's love for books that had been read only by the Venerable Bede, urged him with a certain fatherly malice to get into the discussion, and without even taking a breath, he explained that the cockroach, the

267

oldest winged insect on the face of the earth, had already been the victim of slippers in the Old Testament, but that since the species was definitely resistant to any and all methods of extermination, from tomato slices with borax to flour and sugar, and with its one thousand six hundred and three varieties had resisted the most ancient, tenacious, and pitiless persecution that mankind had unleashed against any living thing since the beginning, including man himself, to such an extent that just as an instinct for reproduction was attributed to humankind, so there must have been another one more definite and pressing, which was the instinct to kill cockroaches, and if the latter had succeeded in escaping human ferocity it was because they had taken refuge in the shadows, where they became invulnerable, because of man's congenital fear of the dark, but on the other hand they became susceptible to the glow of noon, so that by the Middle Ages already, and in present times, and *per omnia secula seculorum*, the only effective method for killing cockroaches was the glare of the sun. The encyclopedic coincidence was the beginning of a great friendship. Aureliano continued getting together in the afternoon with the four arguers, whose names were Álvaro, Germán, Alfonso and Gabriel, the first and last friends that he ever had in his life. For a man like him, holed up in written reality, those stormy sessions that began in the bookstore at 6.00 p.m. and ended at dawn in the brothels were a revelation.

That wise Catalan was in fact Ramon Vinyes, the Barranquilla bookseller, cultural activist, and founder of *Voces* magazine (1917–20), first Spanish immigrant, then Spaniard in exile, teacher, dramatist and storyteller. His bookshop, R. Viñas & Co, a pre-eminent cultural centre, was burnt down in 1923 and is still remembered today in Barranquilla as one

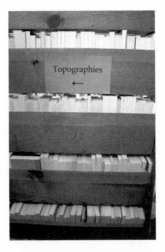

of the mythical bookshops of the Colombian Caribbean. When he went into exile in Latin America as a Republican intellectual after crossing France, he took up teaching and journalism and became the master of a whole young generation known as "the Barranquilla Group" (Alfonso Fuenmayor, Álvaro Cepeda Samudio, Germán Vargas, Alejandro Obregón, Orlando Rivera "Figurita," Julio Mario Santo Domingo and García Márquez). On one of my strangest mornings ever, I gave a taxi driver at the Barranquilla bus station the following address: calle San Blas, between Progreso and 20 de Julio. Librería Mundo. As we drove on, he told me that the names had changed, he did some consulting and discovered that I was referring to calle 35 between Carrera 41 and 43. We headed there. The Librería Mundo run by Jorge Rondón Hederich was where the legendary group of intellectuals met, the spiritual heir of R. Viñas & Co. that had been reduced to ashes twenty years earlier. When I got there, I discovered that it, too, no longer existed. It was obvious, but neither Juan Gabriel Vásquez (who had given me the information) nor I had thought to check it. The bookshop should have been there, but it wasn't, because for quite some time it had only existed in books:

> In any case, the axis of our lives was the Librería Mundo at twelve noon and six in the evening, on the busiest block of calle San Blas. Germán Vargas, an intimate friend of the owner, Don Jorge Rondón, was the one who convinced him to open the store that soon became the meeting place for journalists, young writers and politicians. Rondón lacked business experience, but he soon learned, and with an enthusiasm and a generosity that turned him into an unforgettable Maecenas. Germán, Álvaro and Alfonso were his advisers in ordering books, above all the new books coming from Buenos Aires, where publishers had begun the translation, publication and mass distribution of new literature from all over the world following the Second World War. Thanks to them we could read in a timely way books that otherwise would not have come to the city. The publishers themselves encouraged their patrons and made it possible for Barranquilla to again become the centre of reading it had been years earlier, until Don Ramon's historic bookshop ceased to exist. It was not too long after my arrival when I joined the brotherhood that waited for the travelling salesmen from the Argentinian publishers as if they were envoys from heaven. Thanks to them we were early admirers of Jorge Luis Borges, Julio Cortázar, Felisberto Hernández, and the English and North American novelists who were well translated by Victoria Ocampo's crew. Arturo Barea's *Making of Rebel* was the first hopeful message from a remote Spain silenced by two wars.

That is García Márquez writing about those two bookshops, the one he didn't know and the one he visited, both melded into one in the virtual reality of his masterpiece. I have been unable to find photographs of R. Vinyes & Co. or Mundo on the web and I now realize that this book has found its rhythm in searches inside material books and on the non-material screen, a syntax of to-ing and fro-ing as continuous and discontinuous as life itself; how Montaigne would enjoy the ability of search engines to generate associations, links, fertile byways and analogies. How his heir, Alfonso Reyes, would also have learned from them about whom the narrator of the first part of *The Savage Detectives* says: "Reyes could be my little home. Reading only him and those he liked one could be incredibly happy." *In Books and Bookshops in Antiquity* the erudite Mexican noted:

> Parchment was cheaper and more resistant than papyrus, but the book trade did not adopt it as a matter of course[. . .] Ancient producers of books preferred this light, elegant material, and there was a degree of aversion towards the weight and coarseness of parchment. Galen, the great doctor from the second century AD, was of the opinion that, for reasons of hygiene, shiny parchment hurt and tired the eyes more than smooth opaque papyrus that did not reflect the light. Ulpianus the jurist (died AD 229) examined as a legal problem the issue of whether codices made of vellum or parchment should be considered as books in library bequests, something that did not even have to be debated in the case of papyrus items.

Almost two millennia later, the slow transition from reading paper to reading onscreen gives these arguments a contemporary twist. We now wonder if the screen and the light it radiates do more damage to the eyes than electronic ink, which does not allow us to read in the dark. Or whether, after someone's death, it is right for their heirs to inherit, through books, vinyl records,

CDs and hard discs, the songs and texts their parents bought for themselves. Or whether television and video games harm the imagination of children or adolescents, because they stimulate their reflexes but damage the activity of their brains and are so violent. As Roger Chartier has studied in *Inscription and Erasure, Written Culture and Literature from the Eleventh to the Eighteenth Century*, it is in Golden Age Castile that the danger fiction represents for the reader is first formally expressed, with *Don Quixote* arousing the greatest social fear: "In the eighteenth century, the discourse is medicalized and constructs a pathology of excessive reading that is thought to be an individual sickness or collective epidemic." In this period the reader's sickness is related both to the arousal of the imagination and the immobilizing of the body: the threat is as mental as it is physiological. Following this thread, Chartier analyses the eighteenth-century debate over traditional reading that was called *intensive*, and modern reading that was said to be *extensive*:

> According to this dichotomy, suggested by Rolf Engelsing, the *intensive* reader was confronted by a restricted range of texts that were read and reread, memorized and recited, listened to and learnt by heart, transmitted from generation to generation. Such a way of reading was heavily impregnated with sacred purpose, and subjected the reader to the authority of the text. The *extensive* reader, who appears in the second half of the eighteenth century, is very different and reads countless new, ephemeral printed works and devours them eagerly and quickly. His glance is distanced and critical. In this way, a communitarian, respectful relationship is replaced by irreverent, self-assured reading.

Our way of reading, inextricably linked to screens and keyboards, must be about the spread, books having been produced at an ever-increasing rate, of more and more audio-visual information and knowledge platforms, of that *broadening out,* with all its political implications. The loss of the ability to concentrate on a single text brings the gain of a glimmer of light, critical, ironic distance, the ability to relate and interpret simultaneous phenomena. Consequently, it brings an emancipation from authorities that restrict the range of reading, the deconsecration of an activity that by this stage in evolution should be almost *natural*: reading is like walking, like breathing, something we do without even having to think.

Whilst the apocalyptically minded revamped worn-out arguments from worlds that no longer existed rather than accepting perpetual change as the immutable engine of History, Fnac bookshops filled up on video games and television series and prestigious bookshops began to sell commentaries about video games and television series, as well as eReaders and eBooks. Because the moment a style ceases to be a fashion or trend and becomes mainstream, it will probably undergo a process of sophistication and end up on bookshop and library shelves and in museum rooms. As a cultural product. As a work of art. As a commodity. Scorn of emerging and mainstream styles is fairly common in the world of culture, a field—as they all are—dominated by fashion, the ego and the economy. Most of the bookshops I have mentioned in this essay, on the international circuit where I have slotted myself in as tourist and traveller, nurture a class fiction to which greater millions now have access—fortunately—but they are still a minority. We represent the broadening out of the *chosen people* that Goethe met in the Italian bookshop. A class fiction that is eminently economic—as they all are—even though it wears a veneer of an education that is more or less refined. We should not deceive ourselves: bookshops are cultural centres, myths, spaces for conversations and debate, friendships and even amorous encounters, due in part to their pseudo-romantic paraphernalia, which is often championed by readers who love their craftsmanship, and even by intellectuals, publishers and writers who know they form part of the history of culture. But above all bookshops are businesses. And their owners, often charismatic booksellers, are also bosses, responsible for paying the wages of their employees and ensuring their labour rights are respected, managers, overseers, negotiators skilled in the ins and outs of labour legislation. One of the most inspiring and sincere pieces of those gathered together in *Rue de l'Odéon* is in fact the one that links freedom to the purchase of a book:

> For us the business has a deep and very moving meaning. In our view a shop is a real magic chamber: when a passer-by crosses the threshold of a door that anyone can open, enters this impersonal place, one might say that nothing changes the expression on his face or his tone of voice: with a feeling of total freedom he is carrying out an act he believes has no unexpected consequences.

But which in fact is defined by those consequences: James Boswell will meet Samuel Johnson in Tom Davies' bookshop on Russell Street; Joyce will find a publisher for *Ulysses*; Ferlinghetti will decide to open his own bookshop in San Francisco; Josep

Pla will enter the Canet bookshop in Figueras as a child and seal his pact with literature; William Faulkner will work in one as a bookseller; Vargas Llosa will buy *Madame Bovary* in a bookshop in the Quartier Latin in Paris a long time after seeing the film in Lima; Jane Bowles will meet her best friend in Tangier; Jorge Camacho will buy *Singing from the Well* in a bookshop in Havana and become Reinaldo Arenas' main champion in France; a psychiatrist will advise a juvenile delinquent by the name of Limi-nov to go to Bookshop 41 in a provincial Russian city and this will make a writer of him; François Truffaut will find a novel by Henri-Pierre Roché entitled *Jules et Jim* among second-hand books in Delamaine in Paris; one night in 1976 Bolaño will read the "First Infra-realist Manifesto" in Ghandi Bookshop in Mexico City; Cortázar will discover Cocteau's work; Vila-Matas will find Borges. Perhaps it was only once that the fact somebody did not enter a bookshop had positive outcomes: one day in 1923, Akira Kurosawa headed off to Tokyo's famous Maruzen bookshop, renowned because of its building constructed by Riki Sano in 1909 and for importing international titles for the Japanese cultural elite. He was planning to buy a book for his sister, but he found that the shop was shut and left; two hours later, an earthquake destroyed the building and the whole dis-trict was consumed by flames. Literature is magic and exchange, and for centuries has been sustained, like money, by paper, which is why it has fallen victim to so many fires. Bookshops are busi-nesses on two simultaneous, inseparable levels: the economic and the symbolic, the sale of copies and the creation and destruc-tion of reputations, the reaffirmation of dominant taste or the invention of a new one, stocks and credits. Bookshops have always been the canon's witches' Sabbath and hence key points in cultural geopolitics. The places where culture becomes more physical and thus more open to manipulation. The spaces where, from district to district, town to town, city to city, it is decided what reading matter people will have access to, what is going to be distributed and thus open to the possibility of being consumed, thrown away, recycled, copied, plagiarized, parodied, admired, adapted or translated. It is where their degree of *influence* is mainly decided. It was not for nothing that the first title Diderot gave to his *Letter on the Book Trade* was: "A political and historical letter written to a magistrate about the Bookshop, its present and ancient status, its rules, its privileges, its tacit limits, the censors, itinerant sellers, the crossing of bridges and other matters related to the control of literature."

The Internet is changing that democracy—or dictatorship, depending on how you look at it—of distribution and selection. I often buy titles published in cities I have visited and was unable to buy when I was there from Amazon or other web stores. Last year, on my return from Mexico City, where I *exhausted* a dozen bookshops looking for an essay by Luis Felipe Fabre published by a small Mexican publishing house, I decided to look on the Casa del Libro page and there it was and cheaper than in its place of origin. If Google is the Search Engine and Barnes & Noble the Book Chain, it hardly needs to be said that Amazon is the supreme Virtual Bookshop. Though that is not very precise: even if it was born in 1994 as a bookshop with the name of Cadabra.com and soon after switched to Amazon in order to shoot up the alphabetical pecking order that ruled the Internet before Google, the truth is that for some time it has been a big department store where books are as important as cameras, toys, shoes, computers or bicycles, although the brand bases its power to pull in customers on emblematic devices like the Kindle, a *reader* or electronic book that creates customer loyalty. Indeed, in 1997 Barnes & Noble took it to court over its deceitful advertising (that tautology): the slogan "The world's greatest bookstore" was not true because Amazon was a *book broker* and not a *bookstore*. Now it deals in anything that is on offer, except for eReaders that are not Kindles.

We are innate searchers of the physical world—my hunt for the non-existent bookshop in Barranquilla is only one example from a thousand—and cannot stop being that in the virtual world as well: the history of the electronic book is as gripping as a thriller. It began in the 1940s, gathered speed in the 1960s with hypertext publishing and found a format in the 1970s thanks to Michael S. Hart and a description ("electronic book") thanks to Professor Andries Van Damme, of Brown University, in the middle of the 1980s. When Sony launched its book reader in 1992 with the Data Discman CD, it did so with the tag "The library of the future." Kim Blagg got the first ISBN for an electronic book in 1998. These are the data, the possible chronologies, the clues that, when combined, create the feeling that we are caught between two worlds, as were Cervantes' contemporaries in the seventeenth century, Stefan Zweig's at the beginning of the twentieth or the inhabitants of Eastern Europe the end of the 1980s. In a slow apocalypse in which bookshops are at once oracles and privileged observatories, battlefields and twilight horizons in an irrevocable process of mutation. As Alessandro Baricco says in *The Barbarians*:

> It is a mutation. Something that concerns everyone, without exception. Even the engineers, up there on the wall's turrets are already starting to take on the physical features of the very nomads they, in theory, are fighting against and they have nomadic coins in their pockets, as well as dust from the steppes on their starched collars. It's a mutation. Not some minor change or inexplicable degeneration, or mysterious disease, but a mutation undergone for the sake of survival. The collective choice of a different, salutary habitat. Do we have even the vaguest sense of what could have generated it? I can certainly think of a number of decisive technological innovations, the ones that have compressed space and time, squeezing the world. But these probably would not have been enough had they not coincided with an event that threw open the whole social scene: the collapse of the barriers that until now had kept a good part of humanity far from the routines of desire and consumption.

The word desire reappears yet again in this book, that chemical energy that draws us to certain bodies and objects, vehicles towards manifold knowledge. In the post-1991 world, with neo-liberalism strengthened by the fall of the Soviet Union, and increasingly digital and digitized, that desire has been assuming material form in the consumption of the pixel, that smallest unit of information with which we make sense of our writing, photographs, conversations, videos and maps that explain the routes where we sweat, drive, fly or read. That is why bookshops have web pages: in order to sell us pixelated books, and so we also consume images, stories, the latest novelties and gimmicks. All this is substantial, not mere accident: our brains are changing, the way we communicate and relate is changing: we are the same but very different. As Baricco explains, in recent decades, what we understand by experience and even the tissue of our existence has changed. The consequences of this mutation are as follows: "Surface rather than depth, speed rather than reflection, sequences rather than analysis, surfing rather than penetration, communication rather than expression, multitasking rather than specialization, pleasure rather than effort."

An exhaustive dismantling of the machinery of nineteenth-century bourgeois thought, a final destruction of the last debris of the shipwreck of the divine in everyday life. The political victory of irony over the sacred. It is much more difficult for the few gods of old that survived two world wars on paper to continue to harass us from the dull glow of the screen.

Cultures cannot exist without memory, but need forgetfulness too. While the Library insists on remembering everything, the Bookshop selects, discards, adapts to the present thanks to a necessary forgetfulness. The future is built on obsolescence; we have to discard past beliefs that are false or have become obsolete, fictions and discourses that do not shed the faintest light. As Peter Burke has written: "Discarding knowledge in this way may be desirable or even necessary, at least to an extent, but we should not forget the losses as well as the gains." That is why once the inevitable process of selecting and discarding has taken place, one should "study what has been dispensed with over the centuries, the intellectual refuse," where humanity might have got it wrong, where what was most valuable might have been cast into oblivion, among data and beliefs that did deserve to disappear. After so many centuries of long-term survival, books, due to electronic sourcing, are entering into the logic of inbuilt obsolescence, of a sell-by date. This will bring an even more profound change to our relationship with the texts we are going to be able to translate, alter and *personalize* to an unimaginable degree. It is the crossroads on the journey that began with humanism when philology questioned useless, hackneyed authorities and Bibles began to be turned into our languages via rational criteria and not according to the say-so of superstition.

If there are still many of us who keep collecting futile stamps on our foolish passports to the bookshops of the world it is because we find there the remains of cultural gods that have replaced the religious sort. From Romantic times to the present, like archaeological ruins, like some cafés and so many libraries, or cinemas and museums of contemporary art, bookshops have been and still are ritual spaces, often marked out by tourism and other institutions as ways to understand the history of culture, erotic topographies, and stimulating contexts to find material to nourish our place in the world. If with the death of Jakob Mendel or the hypertext of Borges those physical places we can cling to became more fragile and less transcendent, with the Internet they are much more virtual than our imagination might suggest. They compel us to create new mental tools, to read more critically and more politically than ever, to imagine and connect as never before, analyzing and surfing, going deeper and more rapidly, transforming the privilege of unheard-of access to Information into new forms of Knowledge.

I devote many of my Sunday afternoons to surfing the web in search of bookshops that still do not exist, though they are out there, waiting for me. For years, I have been a reader-viewer of emblematic places I have yet to visit. Very recently, chance enabled me to get to know two of them: in Coral Gables, whose name had always evoked Juan Ramón Jiménez, twenty-four hours of unexpected stopover allowed me to go to Books & Books, a very beautiful Miami bookshop housed in a Mediterranean-style building from the 1920s. One weekend in Buenos Aires when I had nothing planned, I decided to take the ferry and visit Montevideo to finally discover in person an even more beautiful, equally well-stocked bookshop, Más Puro Verso, with its art deco architecture from the same era and glass display cabinet at the top of its imperial stairs. Just as I coveted those spaces, I have spent years collecting leads to others in books, magazines, web pages or videos. For instance, Tropismes in a nineteenth-century arcade in Brussels; Les Bals des Ardents in Lyons with that grand door made from books and Oriental carpets that invite one to read on the floor; Bordeaux's Mollat, which has just turned every book-lover's dream into a reality: the chance to spend a night in a bookshop, and whose website is always bubbling with ideas and activities, a wholly family tradition transformed into 2,500 metres of printed culture overflowing from the very same house where no less a figure than the traveller-philosopher Montesquieu lived, wrote and read at the beginning of the eighteenth century; Candide, its architecture as light as bamboo, in Bangkok, run by the writer, publisher and activist Duangruethai Esanasatang; Athenaeum Boekhandel in Amsterdam, which Cees Nooteboom emphatically recommended to me for its classical aesthetics and, above all, for its importance as a cultural centre and writers' residence; Pendleburys, a country house devoured by a Welsh forest; Swipe Design in Toronto, because an antique bicycle hangs from its ceiling and a chessboard sits between its two readers' armchairs;

Ram Advani Booksellers, the mythical shop in Lucknow, although now I will not be able to meet Ram Advani, who died at the end of 2015 at the age of ninety-four, and whose memory is perpetuated by his daughter-in-law, Anuradha Roy; and Atomic Books, the favourite bookshop of Santiago García the scriptwriter and comics critic who in an email told me that it is one of the best in the US for a reader of graphic novels, though they also sell literature, countercultural fanzines and even toys and punk records: "What's more, you can meet John Waters picking up his mail." I have no information about the history or importance of others, photographs have simply captivated me, because everything I have about them is in languages like Japanese, which I do not understand: Orion Papyrus, in Tokyo, with its parquet floors, its lights worthy of Mondrian and that blend of wood and metal in shelves full of art and design books, or Shibuya Publishing & Booksellers in the same city, with bookshelves in every imaginable geometric shape.

And if I ever return to Guatemala City, I will fight against my nostalgia for El Pensativo, which has disappeared, and will *religiously* repair to Sophos. I expect I will jot down notes on them all when I pay them a visit, like someone who is paying off their debts, in a notebook similar to the one I used on that far-off trip, because I have now given up on my iPad's Moleskine app and do not like my mobile phone doubling as a camera and a notebook. You see: what matters in the end is the will to remember.

In "Covert Joy," a story by Clarice Lispector, we meet a girl who was "fat, short, freckled and had reddish, excessively frizzy hair" but who had "what any child devourer of stories would wish for: a father who owned a bookshop." Many years ago I started to peel off the sticker with the price and barcode on any book I bought and stick them on the inside of the back cover next to the anti-theft chip. It was my way of maintaining an almost fatherly link. The last wish of writer David Markson, who died in New

York in June 2012, was for his library to be sold in its entirety to the Strand and thus be scattered among all those many, many libraries of innumerable anonymous readers. For one dollar, or twenty, or fifty, his books went there, were reintegrated into the market where they once belonged to await their fate and fortune. Markson could have bequeathed his library to a university, where it would have accumulated dust and been visited by the few specializing in his work, but he opted for the opposite move: to share it around, break it up and subject it to the risk of totally unexpected future readings. When the news broke, dozens of the followers of the author of *This is Not a Novel* rushed to the Manhattan bookshop to locate his annotated, underlined books. A virtual group was set up. Scanned pages started to be published on the Internet. In his copy of *Bartleby the Scrivener*, Markson underlined every appearance of the phrase "I would prefer not to"; in *White Noise*, he alternated "astonishing, astonishing, astonishing" with "boring, boring, boring"; in a biography of Pasternak he wrote in the margin: "It is a fact that Isaak Babel was executed in the basement of a Moscow prison. A very strong possibility that the manuscript of an unpublished novel is still around in Stalin's archives." One could turn all the marginal comments in Markson's library into one of his fragmentary novels, where notes on reading, poetic impressions and reflections follow on as if it were a zapping session. It would be an impossible novel because nobody is ever going to find all the books that made up his library: many of them were bought or are being bought by people who do not know who Markson was. That gesture forms part of his legacy. A final, definitive gesture combining death, inheritance, paternity and a single one of the infinite bookshops that sum up all the rest, a unique story dedicated to world literature.

Ideas only exist in things.
David Markson, *The Loneliness of the Reader*

WEBOGRAPHY

American Booksellers Association: http://www.bookweb.org
Bloc de Llibreries: http://www.delibrerias.blogspot.com.es
Book Forum: http://www.bookforum.com
Book Mania: http://www.bookmania.me
Bookseller and Publisher: http://www.booksellerandpublisher.com.au/
Bookshop Blog: http://www.bookshopblog.com
Books Live: http://www.bookslive.co.z
Bookstore Guide: http://www.bookstoreguide.org
Book Patrol: http://www.bookpatrol.net
Courrier du Maroc: http://www.courrierdumarroc.com
Día del Libro: http://www.diadellibro.eu
Diari d'un llibre vell: http://www.llibrevell.cat
El Bibliómano: http//www.bibliographos.net
El Llibreter: http://www.llibreter.blogspot.com.es/
El Pececillo de Plata: http://www.elpececillodeplata.wordpress.com/
Gapers Block: http://www.gapersblock.com
José Luis Checa Cremades. Bibliofilia y encuadernación: http://www.
checacremades.blogspot.com.es
Histoire du Livre: http://www.histoire-du-livre.blogspot.com.es
Kipling: http://www.kipling.org.uk
Le Bibliomane Moderne: http://www.le-bibliomane.blogspot.com.es
Libbys Book Blog: http://www.libbysbooksblog.blogspot.com.es
Library Thing: http://www.librarything.com
Libreriamo: http://www.libreriamo.it
Paul Bowles Official Site: http://www.paulbowles.org
Rafael Ramón Castellanos Villegas: http://www.rrcastellanos.blogspot.
com.es
Reading David Markson: http://www.readingmarksonreading.tumblr.
com
Rare Books Collection de Princeton: http://www.blogs.princeton.edu/
rarebooks/
Reality Studio. A Williams S. Burroughs Community. http://www.
realitystudio.org
Rue des Livres: http://www.rue-des-livres.com

The Bookshop Guide: http://www.inprint.co.uk/thebookguide/shops/
index.php
The Bookseller: http://www.thebookseller.com
The China Beat: http://www.thechinabeat.org
The Haunted Library: http://www.teensleuth.com/hauntedlibrary
The Ticknor Society Blog: http://www.ticknor.org/blog/

FILMOGRAPHY

Before Sunrise (1995), Richard Linklater.
Before Sunset (2004), Richard Linklater.
Chelsea Girls (1966), Andy Warhol and Paul Morrisey.
Fun in Acapulco (1963), Richard Thorpe.
Funny Face (1957), Stanley Donen.
Hugo (2011), Martin Scorsese.
Julie & Julia (2009), Nora Ephron.
Fantômes de Tanger (1997), Edgardo Cozarinsky.
The Life of Others (2006), Florian Henckel von Donnersmarck.
Lord Jim (1965), Richard Brooks.
9 1/2 Weeks (1986), Adrian Lyne.
Notting Hill (1999), Roger Mitchell.
Portrait of a Bookstore as an Old Man (2005), Benjamin Sutherland and
    Gonzague Pichelin.
Remember Me (2010), Allen Coulter.
Reservoir Dogs (1992), Quentin Tarantino.
Short Circuit (1986), John Badham.
Short Circuit 2 (1988), Kenneth Johnson.
The West Wing (1999–2006), NBC.
Vertigo (1958), Alfred Hitchcock.
You've Got Mail (1998), Nora Ephron.

BIBLIOGRAPHY AND ACKNOWLEDGEMENTS

The author and publisher acknowledge the following sources of copyright material where noted below and are grateful for the permissions granted. Every effort has been made to identify the sources of all the material used. If any errors or omissions are brought to our notice, we will be happy to include the appropriate acknowledgements on reprinting.

Aínsa, Fernando, *Del canon a la periferia: Encuentros y transgresiones en la Literatura uruguaya*, Trilce, Montevideo, 2002.

Barbier, Frédéric, *Histoire du livre*, Armand Colin, Paris, 2006.

Baricco, Alessandro, *The Barbarians: An Essay on the Mutation of Culture*, translated by Stephen Sartarelli, Rizzoli Ex Libris, New York, 2006. © 2006 by Alessandro Baricco, English translation © 2013 by Stephen Sartarelli. Reproduced by permission of Rizzoli/Ex Libris.

Barthes, Roland, *Empire of Signs*, translated by Richard Howard, Jonathan Cape, London, 1983.

Battles, Matthew, *Library: An Unquiet History*, W. W. Norton and Company, New York, 2003.

Bausili, Mercè and Emili Gasch, *Llibreries de Barcelona: Una guia per a lectors curiosos*, Columna, Barcelona, 2008.

Beach, Sylvia, *Shakespeare and Company* Lincoln and London, Nebraska UP, 1980. Copyright © 1959 by Sylvia Beach and renewed 1987 by Frederic Beach Dennis. Used by permission of Houghton Mifflin Harcourt Publishing Company. All rights reserved.

—*The Letters of Sylvia Beach*, edited by Keri Walsh, Columbia University Press, New York, Lincoln, NE, 2010.

Becerra, Juan José, *La interpretación de un libro*, Avinyonet del Penedés, Candaya, 2012.

Bechdel, Alison, *Are You My Mother? A Comic Drama*, Houghton Mifflin Harcourt, New York, 2012.

Benjamin, Walter, *The Arcades Project*, edited by Rolf Tiedemann, translated by Howard Eiland and Kevin McLaughlin, Belknap Press, New York, 2002. Copyright © 1999 by the President and Fellows of Harvard College.

—*One Way Street and Other Writings*, translated by Amit Chaudhuri, Penguin Books, London, 2009.

Bolaño, Roberto, *2666*, translated by Natasha Wimmer, Picador, London, 2009.

—*Between Parentheses (1998–2003)*, translated by Natasha Wimmer, Picador, London 2012. © 2004 by the Heirs of Roberto Bolaño. English translation copyright © 2011 by Natasha Wimmer. Reproduced by permission of Pan Macmillan via PLSclear.

—*Consejos de un discipulo de Morrison a un fanatico de Joyce*, by Roberto Bolaño and Antoni Garcia Porta. Copyright © 1984, Roberto Bolaño y Antoni García Porta.

—"Dance Card," *Last Evenings on Earth,* translated by Chris Andrews, Vintage, London, 2007. Copyright © 1997, 2001, Roberto Bolaño. English translation © Chris Andrews, 2007.

—*The Insufferable Gaucho*, translated by Chris Andrews, Picador, London, 2014.

—*The Savage Detectives*, translated by Natasha Wimmer, Picador, London, 2008. Copyright © 1998, Roberto Bolaño. English translation © Natasha Wimmer 2007, reproduced with permission from Farrar, Straus & Giroux.

—"Vagabond in France and Belgium," *Last Evenings on Earth*, translated by Chris Andrews, Vintage, 2007

Borges, Luis Jorge "Funes the Memorious" and "The Library of Babel," *Ficciones*, translated by Anthony Kerrigan, Grove Atlantic, New York, 1962. Copyright © María Kodama 1998. English translation copyright © 1962 by Grove Press Inc. Used by permission of Grove/Atlantic, Inc. Any third party use of this material outside of this publication is prohibited.

Bourdieu, Pierre, *Distinction: A Social Critique of the Judgement of Taste*, translated by Richard Nice, Routledge, 1984. © 1984, 2010, the President and Fellows of Harvard College and Routledge. Reproduced by permission of Taylor & Francis Books Ltd.

Bowles, Jane, *Out in the World: Selected Letters (1935–70)*, edited by Millicent Dillon, Black Sparrow Press, Santa Barbara, 1986. Copyright © 1985 by Rodrigo Rey Rosa.

Bowles, Paul, *In Touch: The Letters of Paul Bowles*, edited by Jeffrey Miller, Farrar, Straus, Giroux, New York, 1994.

—*Without Stopping*, Putnam, New York, 1972.

—*Travels: Collected Writings, 1950–93*, Sort of Books, London, 2010. Copyright © 2010 by Rodrigo Rey Rosa.

Bradbury, Ray, *Fahrenheit 451*, Rupert Hart-Davis Ltd., London, 1954.

Bridges, Lucas E., *Uttermost Part of the Earth*, Dutton Books, New York, 1949.

Burke, Peter, *A Social History of Knowledge II: From the Encyclopaedia to Wikipedia*, Polity Press, Cambridge, 2012. Copyright © Peter Burke 2012.

Campaña, Mario, *Baudelaire: Juego sin triunfos*, Debate, Barcelona, 2006.

Campbell, James, *This Is The Beat Generation New York – San Francisco – Paris*, Secker & Warburg, London, 1999.

Canetti, Elias, *Auto da Fé*, translated from the German by C. V. Wedgwood, The Harvill Press, London, 2005.

—*The Voices of Marrakesh*, translated by J. A. Underwood, Marion Boyars Publishers, London, 1967.

Carey, Peter, *The True History of the Kelly Gang*, Faber & Faber, London, 2001.

Carpentier, Alejo, *Los pasos recobrados: Ensayos de teoría y crítica literaria*, Biblioteca Ayacucho, Caracas, 2003.

Casalegno Giovanni (Ed.), *Storie di Libri: Amati, misteriosi, maledetti*, Einaudi Editore, Turin, 2011.

Casanova, Pascale, *The World Republic of Letters*, translated by M. B. DeBevoise, Harvard University Press, Cambridge, MA, 2004.

Cavallo, Guglielmo, Roger Chartier and Lydia G. Cochrane, *A History of Reading in the West*, University of Massachusetts Press, Amherst, 2003.

Certeau, Michel de, *Le Lieu de l'Autre: histoire réligieuse et mystique*, Editions du Seuil, Paris, 2005.

Chartier, Roger, *Inscription and Erasure, Written Culture and Literature from the Eleventh to the Eighteenth Century*, Pennsylvania University Press, Philadelphia, PA, 2007.

Chatwin Bruce, *In Patagonia*, Vintage, London, 1998.

—*Under the Sun: The Letters*, edited by Elizabeth Chatwin and Nicholas Shakespeare, Jonathan Cape, London, 2010.

Chih Cheng, Lo, *Bookstore in a Dream*, The Chinese University Press, Hong Kong, 2011.

Choukri, Mohamed, *Paul Bowles in Tangier*, translated by Gretchen Head and John Garret, Telegram, San Francisco, 2008.

Clemente San Román, Yolanda, "Los catálogos de librería de las sociedades Anisson-Posuel y Arnaud-Borde conservados en la Biblioteca Histórica de la Universidad Complutense," *Revista General de Información y Documentación*, vol. 20, 2010.

Cobo Borda, Juan Gustavo, "Libreros colombianos, desde el constitucionalista don Miguel Antonio Caro hasta Karl Buchholz" (http://www.ciudadviva.gov.co/portal/node/32).

Coetzee, J.M., *Disgrace*, Vintage, London, 2000.

—*Dusklands*, Secker and Warburg, London, 1982. Reprinted by permission of The Random House Group Ltd.

Cole, Teju, *Open City*, Faber & Faber, London, 2011.

Cortázar, Julio, "House Taken Over," *Blow Up and Other Stories*,

—*Cartas: 1937–1963*, edited by Aurora Bernádez, Alfaguara, Madrid, 2000. Reproduced with permission from Editorial Alfaguara.

—*Hopscotch*, translated by Gregory Rabassa, Panther, 2014.

Cuadros, Ricardo, "Lo siniestro en el aire" (http://www.ricardo-cuadros.com/html/lo_siniestro.htm).

Dahl, Svend, *A History of the Book*, The Scarecrow Press, New Jersey, 1968.

Debord, Guy, *Commentaries on the Society of the Spectacle*, Verso, London, 1990.

DeMarco, Eileen S., *Reading and Riding: Hachette's Railroad Bookstore Network in Nineteenth Century France*, Associated University Press, Cranbury, 2006.

Diderot, Denis, *Letter on the Book Trade*, translated by Arthur Goldhammer, Daedalus, vol. 131 no. 2: pp 48–56.

Didi-Huberman, Georges, *Confronting Images: Questioning the Ends of a Certain History of Art*, translated by John Goodman, Penn State University Press, 2005.

—*Atlas: How to Carry the World on One's Back*, TF Editores/Museo Nacional Reina Sofia, Madrid, 2010.

Domingos, Manuela D., Bertrand, *Uma livraria antes do Terremoto*, Biblioteca Nacional, Lisbon, 2002.

Donoso, José, *Diarios, ensayos, crónicas: La cocina de la escritura*, edited by Patricia Rubio, Ril Editores, Santiago de Chile, 2008.

Edwards, Jorge, *Persona non grata: A memoir of disenchantment with the Cuban Revolution*, translated by Andrew Hurley, The Paragon Press, New York, 1993.

Eliot, Simon, Andrew Nash and Ian Wilson, *Literary Cultures and the Material Book*, The British Library Publishing Division, London, 2007.

Énard, Mathieu, *Street of Thieves*, translated by Charlotte Mandell, Fitzcarraldo Editions, London, 2015. Copyright © Mathias Énard, 2012. Translation copyright © Charlotte Mandell, 2014. Reproduced by permission of Fitzcarraldo Editions.

Fernández, Benito J., *Eduardo Haro Ibars: los pasos del caído*, Editorial Anagrama, Barcelona, 2005.

Fernández, Eduardo, *Soldados de cerca de un tal Salamina. Grandezas y miserias en la Galaxia Librería*, Comanegra, Barcelona, 2008.

Fernández del Castillo, Francisco, editor, *Libros y libreros en el siglo XVI*, Fondo de Cultura Económica, Mexico City, 1982.

Foucault, Michel, *The Order of Things: An Archaelology of the Human Sciences*, translated by E. Frost, Routledge, 2001.

García Márquez, Gabriel, *One Hundred Years of Solitude*, translated by Gregory Rabassa, Penguin Books, London, 1972. Reprinted by permission of Agencia Literaria Carmen Balcells.

—*Living to Tell the Tale*, translated by Edith Grossman, Penguin Classics, London, 2003.

Gil, Manuel and Joaquín Rodríguez, *El paradigma digital y sostenible del libro,* Madrid, Trama, 2011.

Goethe, J.W. von, *Italian Journey*, translated by W. H. Auden and Elizabeth Mayer, Penguin Classics, London, 1970. English translation Copyright © 1962 by W. H. Auden, renewed. Reprinted by permission of Curtis Brown, Ltd.

Goffman, Ken, *Counterculture Through the Ages: From Abraham to Acid House*, Villard Books, New York, 2004.

Goytisolo, Juan, *Count Julian*, translated by Helen Lane, Serpent's Tail, 1989.

—*A Cock-Eyed Comedy*, translated by Peter Bush, Serpent's Tail, 2002. Reproduced by permission from Serpent's Tail.

—*Forbidden Territory* and *In Realms of Strife*, translated by Peter Bush, Verso, 2003. Reproduced by permissions from Verso Books.

Guerrero Marthineitz, Hugo, "La vuelta a Julio Cortázar en 80 preguntas," from *Julio Cortázar: Confieso que he vivido y otras entrevistas*, Antonio Crespo, editor, Buenos Aires, LC Editor, 1995.

Hanff, Helene, *84, Charing Cross Road*, Avon Books, New York, 1970.

Hemingway, Ernest, *A Moveable Feast*, Vintage, London, 2000.

Hoffman, Jan, "Her Life Is a Real Page-Turner," New York Times, October 12, 2011.

Jenkins, Henry, *Convergence Culture: Where Old and New Media Collide*, New York University Press, New York, 2006.

Johns, Adrian, *The Nature of the Book: Print and Knowledge in the Making*, Chicago University Press, Chicago, 1998.

Kiš, Danilo, *A Tomb for Boris Davidovitch*, translated by Duška Miki-Mitchell, Penguin Classics, London, 1980.

—*The Encyclopedia of the Dead*, translated by Michael Henry Heim, Faber & Faber, London, 1989.

Krishnan, Shekar, "Wheels within wheels," Indian Express, June 17, 1997.

Kubizek, August, *The Young Hitler I Knew*, translated by Lionel Leventhal, Greenhill Books, London, 2006.

Labarre, Albert, *Histoire du livre*, Presses universitaires de France, Paris, 2001.

Laddaga, Reinaldo, *Estética de laboratorio*, Adriana Hidalgo Editora, Buenos Aires, 2010.

Lernout, Geert and Wim Van Mierlo, *The Reception of James Joyce in Europe Vol.1: Germany, Northern and East Central Europe*, Thoemmes Continuum, London, 2004.

Link, Daniel, "Flaubert & Baudelaire," *Perfil*, Buenos Aires, August 28, 2011.

Lispector, Clarice, "Covert Joy," *The Collected Short Stories*, translated by Katrina Dodson, Penguin Modern Classics, 2015. Copyright ©1951, 1955, 1960, 1978, 2010, 2015 by the Heirs of Clarice Lispector, translation copyright © 2015 by Katrina Dodson. Reprinted by permission of New Directions Publishing Corp. and Penguin Books Ltd.

Loeb Schloss, Carol, *Lucia Joyce: To Dance in the Wake*, Farrar, Straus, Giroux, New York, 2004.

Lovecraft, H. P., "The Battle that Ended the Century," *The Complete Fiction Collection, Vol. III*, Ulwencreutz Media, 2012.

Lyons, Martyn, *Books: A Living History*, J. Paul Getty Museum, Los Angeles, 2011.

Llanas, Manuel, *El libro y la edición en Cataluña: apuntes y esbozos*, Gremi d'Editors de Catalunya, Barcelona, 2004.

MacCannell, Dean, *The Tourist: A New Theory of the Leisure Class*, Shocken Books, New York, 1976.

MacNiven, Ian S., editor, *The Durrell-Miller Letters 1935–1980*, Faber & Faber, 2003. Letter from Henry Miller to Lawrence Durrell copyright © The Beneficeries of the Estate of Henry Miller 2007. Reproduced with permission of Curtis Brown Ltd., London, on behalf of the Beneficeries of the Estate of Henry Miller.

Mallarmé, Stéphane, "The Book: a Spiritual Instrument," translated by Bradford Cook, *Selected Poetry and Prose*, edited by Mary Ann Caws, New Directions, New York, 1982.

Manguel, Alberto, *A History of Reading*, Harper Collins, London, 1996. © Alberto Manguel, c/o Schavelzon Graham Agencia Literaria.

—*The Library at Night*, Yale University Press, London, 2008.

Manzoni, Cecilia, "Ficción de futuro y lucha por el canon en la narrativa de Roberto Bolaño," *Jornadas Homenaje Roberto Bolaño (1953–2003)*, ed. Ramón Férriz González, *Simposio internacional*, ICCI Casa Amèrica Catalunya, Barcelona, 2005.

Marchamalo, Jesús, *Cortázar y los libros*, Fórcola Ediciones, Madrid, 2011.

Markson, David, *Reader's Block*, Dalkey Archive, Champaign, IL, 1996.

Martí Monterde, Antoni, *Poética del Café: Un espacio de la modernidad literaria europea*, Anagrama, Barcelona, 2007.

Martínez López, María Esther, *Jane Bowles y su obra narrativa: ambigüedad moral y búsqueda de una respuesta existencial*, Ediciones de la Universidad de Castilla-La Mancha, Cuenca, 1998.

Martínez Rus, Ana, *"San León Librero": las empresas culturales de Sánchez Cuesta*, Gijón, Trea, 2007.

Melo, Adrián, *El amor de los muchachos: homosexualidad y literature*, Lea, Buenos Aires, 2005.

Mercer, Jeremy, *Time Was Soft There: A Paris Sojourn at Shakespeare and Co.*, St. Martin's Press, New York, 2005.

Michaud, Joseph A., *Booking in Iowa: The Book Trade In and Around Iowa City. A Look Back*, The Bookery and The Press of the Camp Pope Bookshop, Iowa City, 2009.

Mogel, Leonard, *Making It in Book Publishing*, Arco, New York, 1996.

Monnier, Adrienne, *Rue de l'Odéon*, Albin Michel, Paris, 1960.

—*The Very Rich Hours of Adrienne Monnier*, translated, with an introduction and commentary by Richard McDougall, University of Nebraska Press, Lincoln, NE, 1996.

Montaigne, Michel de, *The Complete Essays*, translated by M. A. Screech, Penguin Classics, London, 1991. English translation copyright © M. A. Screech 1987, 1991, 2003. Reproduced by permission of Penguin Ltd.

Montroni, Romano, *Vendere l'anima: Il mestiere del libraio*, GLF Editore, Laterza, Rome, 2006.

Morand, Paul, *Venices*, translated by Euan Cameron, Pushkin Press, London, 2002.

Moretti, Franco, *Atlas of the European Novel 1800–1900*, Verso, London, 1998.

Morgan, Bill, *Beat Generation in New York: A Walking Tour of Jack Kerouac's City*, City Lights Books, San Francisco, 1997.

—*Jack Kerouac and Allen Ginsberg: The Letters*, edited by Bill Morgan and David Stanford, Viking, New York, 2010. Excerpt from Allen Ginsberg © 2011 by Allen Ginsberg, used by permission of The Wylie Agency (UK) Limited.

Muyal, Rachel, *My Years in the Librairie des Colonnes*, Khbar Bladna, Tangier, 2012.

Nancy, Jean Luc, *On the Commerce of Thinking: of Books and Bookstores*, translated by David Mills, Fordham University Press, New York, 2008.

Nooteboom, Cees, *All Souls Day*, translated by Susan Massotty, Picador, London, 2001. © Cees Nooteboom 1998, English translation © Susan Massotty 2001. First published by Atlas Uitgeverij as *Allerzeilen* in 1998. Reproduced by permission of Pan Macmillan via PLSclear.

Ordóñez, Marcos, *Un jardín abandonado por los pájaros*, El Aleph, Barcelona, 2013.

Osorgin, Mikhail, Alexei Remizov and Marina Tsvetaeva, *La Librería de los Escritores*, translated by Selma Ancira, Edicions de La Central /Sexto Piso, 2007.

Ortiz, Renato, *Modernidad y espacio: Benjamin en París*, translated by María Eugenia Contursi and Fabiola Ferro, Norma, Buenos Aires, 2000.

Otlet, Paul, *Traité de Documentation: le libre sur le libre, théorie et pratique*, Editions Mundaneum, Brussels, 1934.

Palmquist, Peter, and Thomas Kailbourn, *Pioneer Photographers of the Far West: A Biographical Dictionary, 1840–1865*, Stanford UP, 2006.

Parish, Nina, Henri Michaux, *Experimentation with Signs*, Rodopi, Amsterdam, 2007.

Pascual, Carlos, Paco Puche and Antonio Rivero, *Memoria de la Librería*, Trama Editorial, Madrid, 2012.

Paz, Octavio, *The Monkey Grammarian*, translated by Helen R. Lane, Peter Owen Publishers, London, 1989.

Petroski, Henry, *The Book on the Bookshelf*, Vintage Books, London, 2000.

Pirandello, Luigi, *Cuentos para un año*, translated by Marilena de Chiara, Nórdica Madrid, 2011.

Ponte, Antonio José, *Un seguidor de Montaigne mira La Habana / Las comidas profundas*, Verbum, Matanzas, 1985.

Primera, Maye, "La librería del exilio cubano cierra sus puertas," El País, April 26, 2013.

Ramírez, Antonio, "Imagining the bookshop of the future," Huffington Post, September 18, 2012.

Reyes, Alfonso, and Pedro Henríquez Ureña, *Correspondencia 1907–1914*, edited by José Luis Martínez, Fondo de Cultura Económica, Mexico City, 1986.

Rice, Ronald, *My Bookstore: Writers Celebrate Their Favorite Places to Browse, Read, and Shop*, Black Dog & Leventhal Publishers, New York, 2012. Excerpt from "Green Apple Books, San Francisco. CA" by Dave Eggers. Copyright © 2012 Dave Eggers, used by permission of The Wylie Agency (UK) Limited.

Rushdie, Salman, *Joseph Anton: A Memoir*, Jonathan Cape, London, 2012. Copyright © Salman Rushdie 2012. Reproduced by permission of The Random House Group Ltd.

Saint Phalle, Nathalie, *Hoteles literarios: Viaje alrededor de la Tierra*, translated by Esther Benítez, Alfaguara, Madrid, 1993.

Sansieviero, Chachi, "La librería limeña El Virrey," Cuadernos Hispanoamericanos, n. 691, December 2008.

Schiffrin, André, *Words and Money*, Verso Books, London, 2010.

Scott, Anne, *18 Bookshops*, Sandstone Press, Dingwall, 2013.

Sebald, W. G., *The Rings of Saturn*, translated by Michael Hulse, Vintage Books, London, 2002.

—*Austerlitz*, translated by Anthea Bell, (Hamish Hamilton 2001, Penguin Books 2002), London 2002. Copyright © The Estate of W.G. Sebald 2001. English translation © Anthea Bell 2001. Reproduced by permission of Penguin Books Ltd.

Sennett, Richard, *The Craftsman*, Penguin Books, London, 2009. Copyright © Richard Sennett, 2008. Reproduced by permission of Penguin Books Ltd.

Serra, Cristóbal, editor, *Apocalipsis*, Siruela, Madrid, 2003.

Service, Robert, *Lenin: A Biography*, Macmillan, London, 2000. © Robert Service 2000. Reproduced with permission of Pan Macmillan via PLSclear.

—*Stalin: A Biography*, Macmillan, London, 2004.

Shakespeare, Nicholas, *Bruce Chatwin*, Vintage, London, 2000.

Smith, Gibbs M., *The Art of the Bookstore: The Bookstore Paintings of Gibbs M. Smith*, Gibbs Smith, Layton, UT, 2009.

Sontag, Susan, *I, etcetera*, Farrar, Straus, Giroux, New York, 1978.

Sorrel, Patricia and Fréderique Leblanc, *Histoire de la librairie française*, Paris, Editions du Cercle de la Librairie, 2008.

Steiner, George, *Extraterritorial: Papers on language and the language revolution*, Athenaeum, New York, 1971. Copyright © 1968, 1969, 1970, 1971 by George Steiner. Reprinted by permission of Georges Borchardt, Inc., on behalf of the author.

Steloff, Frances, *In Touch with Genius: Memoirs of a Bookseller*, Marian Seldes, Direct Cinema Ltd, Los Angeles, 1987.

Sterne, Laurence, *A Sentimental Journey Through France and Italy*, Lanham Start Classics, 2014.

Talese, Gay, *A Writer's Life*, Knopf, New York, 2006. Copyright © 2006 by Gay Talese. Reproduced by permission of Random House Ltd.

Thorpe Nicholson, Joyce, and Daniel Wrixon Thorpe, *A Life of Books. The Story of D. W. Thorpe PTY LTD 1921–1987*, Courtyard Press, Middle Park, 2000.

Unwin, Sir Stanley, *The Truth about Publishing*, Macmillan, London, 1960.

Verne, Jules, *The Lighthouse at the End of the World*, translated by William Butcher, Nebraska University Press, 2007.

Vila-Matas, Enrique, *Never Any End to Paris*, translated by Anne McLean, Harvill Secker, London, 2014. Copyright © Enrique Vila-Matas 2003. English Translation copyright © Anne McLean, 2011. Reproduced by permission of The Random House Group Ltd.

Vitkine, Antoine, *"Mein Kampf": histoire d'un livre*, Flammarion, Paris, 2009.

Various authors, *El libro de los libros. Guía de librerías de la ciudad de Buenos Aires*, Asunto Impreso Ediciones, Buenos Aires, 2009.

Various authors, *Kerouac en la carretera. Sobre el rollo mecanografiado original y la generación beat*, translated by Antonio Prometeo Moya, Anagrama, Barcelona, 2010.

Various authors, *El origen del narrador: Actas completas de los juicios a Baudelaire y Flaubert*, Mardulce, Buenos Aires, 2011.

Vollmann, William T., *The Royal Family*, New York, Penguin Books, New York 2000. Copyright © William T. Vollmann, 2000.

—*Central Europe*, Alma Books, 2006.

Walser, Robert, *The Walk and Other Stories*, translated by Christopher Middleton, Serpent's Tail, 2013.

Weiss, Jason, *The Lights of Home: A Century of Latin American Writers in Paris*, Routledge, New York, 2003.

Whitman, George, *The Rag and Bone Shop of the Heart*, Shakespeare and Company, Paris, 2000.

Williamson, Edwin, *Borges: A Life*, Viking, New York, 2004.

Yánover, Héctor, *Memorias de un librero*, Anaya & Mario Muchnik, Buenos Aires, 1994.

—*El regreso del Librero Establecido*, Taller de Mario Muchnik, Madrid, 2003.

Zweig, Stefan, *The World of Yesterday*, translated by Anthea Bell, Pushkin Press, London, 2009.

—"Mendel the Bibliophile," *The Collected Stories of Stefan Zweig*, translated by Anthea Bell, Pushkin Press, London, 2013. Translation © Anthea Bell, reproduced by permission from Pushkin Press.